Fluoride

Edited by Enos Wamalwa Wambu,
Grace J. Lagat and Ayabei Kiplagat

Published in London, United Kingdom

Fluoride
http://dx.doi.org/10.5772/intechopen.98000
Edited by Enos Wamalwa Wambu, Grace J. Lagat and Ayabei Kiplagat

Contributors
Thangapandiyan Shanmugam, Miltonprabu Selvaraj, Alexander Dyachenko, Eugenio Hernan Otal, Mutsumi Kimura, Manuela Leticia Kim, Enos Wamalwa Wambu, Franco Frau, Revocatus Machunda, Stephen S. Barasa, Lilliane Pasape, Giorgio Ghiglieri, Babu Rao Gudipudi, Abdulmohsen Alamry

First published in London, United Kingdom, 2022 by IntechOpen
IntechOpen is the global imprint of INTECHOPEN LIMITED, registered in England and Wales, registration number: 11086078, 5 Princes Gate Court, London, SW7 2QJ, United Kingdom

British Library Cataloguing-in-Publication Data
A catalogue record for this book is available from the British Library

Additional hard and PDF copies can be obtained from orders@intechopen.com

Fluoride
Edited by Enos Wamalwa Wambu, Grace J. Lagat and Ayabei Kiplagat
p. cm.
Print ISBN 978-1-80355-642-0
Online ISBN 978-1-80355-643-7
eBook (PDF) ISBN 978-1-80355-644-4

Meet the editors

Dr. Enos W. Wambu is the current Head of the Department of Chemistry and Biochemistry at the University of Eldoret, Kenya, where he has taught Environmental Chemistry since 2014. He holds a Ph.D. in Chemistry from Kenyatta University in Kenya. He has previously taught Physical Chemistry at Jaramogi Oginga Odinga University of Science and Technology (2011 – 2014). Until then, he had been a long-serving, highly successful Graduate Teacher of Mathematics and Chemistry having served in some of the renowned schools in the nation including Friends' High School Kamusinga, St. Lukes Kimili Boys' High School, and Kimugui Secondary School in Bungoma County, Kenya. Dr. Wambu has distinguished himself as a fluoride science specialist and published extensively on this subject since 2009.

Dr. Grace Lagat is a Lecturer and Head of Analytical Chemistry Programmes at the Department of Chemistry and Biochemistry, at the University of Eldoret, Kenya. She holds a Ph.D. (Chemistry) from Kansas University in the USA and an M.Sc. (Education Administration) from the University of Eastern Africa, Baraton, Kenya. She has over 20 years of university teaching experience including a "KUWANA, Graduate Opportunity Academic Scholarship" at the University of Kansas, Kansas, USA. Also winning the "Student and Young Investigator Travel Award" at the 3rd Biennial National Institutional Development Award Symposium of Biomedical Research Excellence, Bethesda, Maryland, USA, in 2010. Her research interests are in the structure-function analysis of biomolecules and their complexes and probing protein-protein interaction hotspots for structure-based drug design.

Dr. Kiplagat Ayabei is a lecturer in the Analytical Chemistry Department of Chemistry and Biochemistry at the University of Eldoret. A native of Kenya, he received B.Sc. in Chemistry from Jomo Kenyatta University of Agriculture and Technology, an M.Sc. (Analytical Chemistry) from the University of Eldoret, and a Ph.D. in Chemistry from the University of the Western Cape, South Africa. His research activities include water quality, water purification, green nanotechnology, and synthesis of nanomaterials for biological applications. He has taught courses in general chemistry, analytical chemistry and Instrumentation.

Contents

Preface

The subject of fluoride has been studied extensively by scientists around the globe. The current interest in fluoride science has, however, arisen in the context of the ongoing climate change scenarios that have aggravated fluoride pollution of natural resources, especially surface and groundwater resources and agricultural soils, with far-reaching impacts on ecosystems, public health, and water and food security.

Although significant gaps remain in the primary data on the current topic, large volumes of research data are continuously being added to the knowledge of environmental fluoride. There is a growing desire among the scientific community to consolidate the present knowledge with a view to improving a broad understanding of the global fluoride problem. A number of review papers have been published on fluoride environmental pollution, its public health impacts, and remediation and amelioration strategies. The present volume aims to contribute to the consolidation of this knowledge and improve its accessibility.

The book is therefore based on the case-study model, aiming to present, logically and concisely, selected works on various aspects of the subject of fluoride, including its occurrence and environmental sources, natural contamination, human exposure, its impacts, and their amelioration. It is hoped that readers will find this volume useful for enriching and advancing their understanding of the subject.

Enos Wamalwa Wambu, Grace J. Lagat and Ayabei Kiplagat
Department of Chemistry and Biochemistry,
University of Eldoret,
Eldoret, Kenya

Chapter 1

Origin and Hydrogeochemistry of Fluoride in the Context of the Yemen Regime

Abdulmohsen Alamry

Abstract

Groundwater is a natural resource that is used in a variety of fields, which has an impact on its quality. In many places of the world, fluoride-enriched water has become a major public health concern. It is necessary to investigate the geochemical mechanism of fluoride enrichment in drinking water. In Yemen, groundwater is the only supply of water, and its quality is critical because it determines the groundwater's usefulness for drinking and other domestic purposes. The primary goal of this chapter is to gain a better understanding of factors that influence high fluoride levels in groundwater and its impacts from selected parts of Yemen. The elevated ion concentrations in groundwater are most likely due to water-rock interaction, according to the regional hydrogeochemical investigation. The main findings of this review indicate that the children in the area who get their drinking water from wells with high fluoride levels are suffering from dental and skeletal fluorosis. The population in the research area is at high risk due to excessive fluoride intake, particularly in the absence of knowledge about quantity of fluoride consumption.

Keywords: hydrogeochemistry, fluorosis, fluoride contamination, volcanic rocks, rock-water interaction, fluorite, Yemen

1. Introduction

Groundwater's chemical composition is obtained from a variety of sources of solutes, including gases and atmospheric aerosols, below-surface replacement and precipitation reactions, weathering and erosional activities of soils and rocks, and other anthropogenic effects. The study of water chemistry can reveal a lot about the geological history of rocks as well as the velocity and direction of water flow [1]. Groundwater's chemical, physical, and bacteriological properties determine its suitability for municipal, commercial, industrial, agricultural, and domestic use [2].

To understand an aquifer's hydrogeochemistry, a detailed understanding of the rock-water interactions that influence groundwater chemical composition is required. The mineral composition of the rock is the primary component that governs a location's water chemistry [3]. The local regime differs from other sites due to the continual interfacial reactivity of water with rocks.

Fluoride, currently considered a pollutant in several regions of the world, is frequently related with the dissolution of fluorine-containing minerals in rocks, as well as growing anthropogenic influences [4]. Groundwater chemistry is influenced by mineral water interfacial interactions such as carbonate weathering and dissolution, silicate weathering, and ion exchange activities. Groundwater composition in shallow alluvial aquifers is controlled by hydrogeochemical processes such as dissolution, cation exchange processes, calcite equilibrium, and residence period, as well as the flow channel. The hydrogeochemical fluctuations of groundwater from a semiarid sedimentary basin are caused by salt leaching from the surface, ion exchange processes, and residence time [5].

Fluoride ion concentrations in groundwater can alter owing to chemical processes including hydration and hydrolysis, weathering and deposition, ion exchange processes, oxidation and reduction that occur during mineral-water contact [6]. These interactions influenced the mobility of dissolved constituents and altered the pH of groundwater in diverse sites. Fluoride levels were found to be excessive (10 mg/l) in several areas of Yemen [7]. The very alkaline groundwater conditions were thought to be the primary cause of fluorite disintegration.

The primary aim of this chapter is to understand the influence of geochemical processes on fluoride enrichment in groundwater in Yemen regime, as well as its relationship with other major element concentrations and health implications. The primary goal of this chapter is to review and improve understanding of the factors that influence high fluoride levels in groundwater samples.

2. Rock-water interaction

Interactions between ground water and the minerals that make up the aquifer system control major-ion chemistry trends in the aquifer system to a large extent. Mineral dissolution and precipitation, oxidation and reduction, and ion exchange are all important geochemical reactions that can affect solute concentrations in groundwater systems. Evaporation and mixing of water from various sources are examples of additional processes that can affect solute concentrations [8].

Natural causes such as rock-water interactions while flowing are of specific groundwater quality concerns. Fluoride is one of the most common geogenic pollutants found in groundwater [9]. More than 260 million people are thought to be impacted by elevated fluoride levels in drinking water across the world. People in more than 230 districts across 20 states in India are experiencing health problems as a result of elevated fluoride levels in groundwater [10]. Many studies have found that high fluoride concentrations in groundwater are frequently linked to longer residence times in crystalline rocks in arid-semiarid climates with abundant Na-HCO_3 and low Ca, as well as alkaline pH [11].

Geographic Information System (GIS) has been used to map and evaluate groundwater quality all around the world [12–14]. Gibbs [15] used total dissolved solids (TDS) vs. Na/(Na + Ca) and TDS vs. Cl/(CI + HCO_3) to identify rock-water interaction. Minerals of various rock types, such as igneous, metamorphic, and sedimentary, entirely or partially dissolve in water depending on chemical weathering resistance. Chemical weathering resistance is high to extremely high in quartz-cemented sandstone, silt, slate, shale, schist, gneiss, and quartzite. Calcite cemented sandstone, limestone, rock salt, gypsum, marble, and basalt, on the other hand, have low to moderate chemical weathering resistance. Different minerals, such as halite, pyrite, gypsum, dolomite, and calcite, demonstrate good water dissolution as a result of these interactions. Because of their low resistance, olivine, pyroxene, hornblende, and biotite dissolve in water via oxidation-reduction and hydrolysis reactions. Feldspar, quartz, and clay dissolve slowly in groundwater due to their considerable resistance to weathering. Fluorite and fluorapatite, among other minerals, are regarded possible sources of fluoride as a groundwater contaminant [16].

3. Dissolution/precipitation of minerals

Fluoride in groundwater comes primarily from natural sources, with earth's crust containing 0.32% fluoride. Weathering and dissolving of fluoride minerals are the primary controls on its concentration in groundwater, since lithology plays a vital impact in its occurrence [17]. The availability and solubility of fluoride minerals, pH, temperature, anion exchange capacity of aquifer materials, type of geological materials, residence time, porosity, structure, depth, groundwater age, and concentration of carbonates and bicarbonates in water all influence the fluoride contamination of groundwater [18].

The chemical study of groundwater provides insight into the geochemical processes that occur in that area. As previously stated, geological formations regulate water quality when they come into contact with flowing water. Several investigations have found that Na-rich, Ca-poor groundwater with an alkaline pH and high HCO^{-}_{3} can mobilize fluoride from fluoride-rich rock formations, resulting in higher F concentrations in groundwater. Ionic exchange between F and hydroxyl ions in fluorite minerals such as mica, amphiboles, illite, and others may occur at higher pH levels. As a result, the alkaline composition of groundwater promotes fluoride ion desorption and hence increases solubility [19].

4. Fluoride contamination and fluorosis in Yemen

Fluorosis is a major public health problem spot-wise all over the world, including Yemen. Endemic fluorosis has been nearly recognized as a major public health problem in six governorates in Yemen. Since the groundwater forms a major source of drinking water in rural areas, rural populations are facing a major health problem in these governorates. Fluorosis-affected areas in various parts of the country are being discovered on a regular basis. As a result, fluorosis remains an endemic problem in Yemen. Unfortunately, proper fluoride mapping has not been carried in Yemen so as to locate areas with normal, low, or high levels of fluoride. Where, the available report prepared by the General Authority of Rural Water Projects (GARWP) about the increasing fluoride content in groundwater (Between 2000 and 2006) in districts of some governorates (Sana'a, Ibb, Dhamar, Taiz, Al-Dhala, and Raimah) considered to be the first alarming report highlighted the problem of fluoride in Yemen [7].

A systematic study and delineation of fluoride contamination from Taiz and Al-Dahla governorates have been conducted by Alamry [7].

The iso-line contour map of fluoride ion concentration was created using the chemical analyses taken from wells and springs in the selected areas. As demonstrated in **Figure 1**, locations with a high fluoride concentration of more than 1.5 mg/l are labeled as fluoride-contaminated (bold green color).

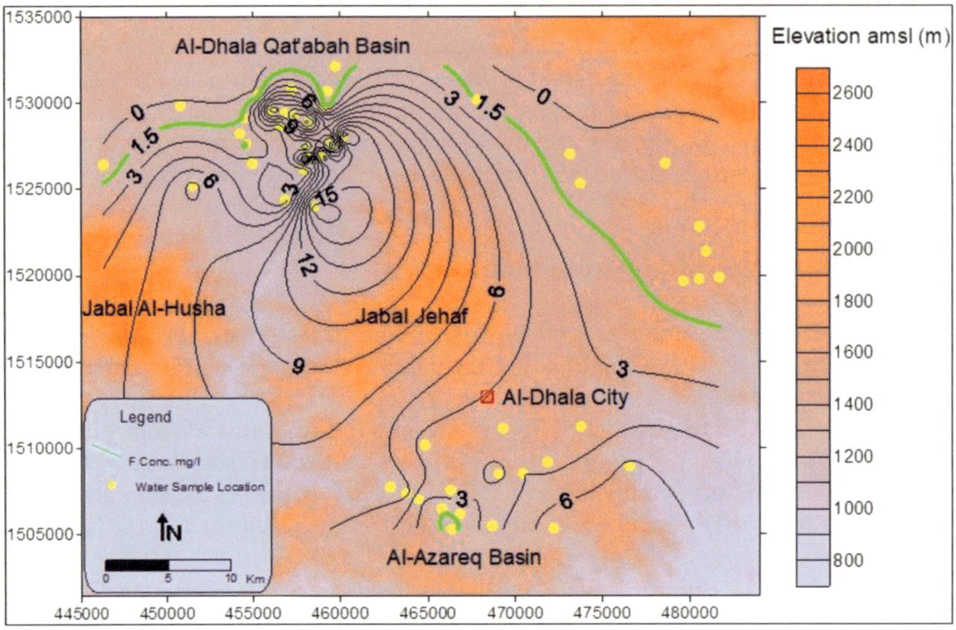

Figure 1.
Iso-line contour map of fluoride ion concentration from Al-Dhala districts.

The fluoride concentrations map clearly shows that there are two significant areas with high fluoride concentrations. These two places are located in the upper portion of Wadi Tuban in the low land area, separated by the Jehaf district's highlands plateau. The first is in the Al-Dhala Qat; abah Basin, stretching from Qarad to Qa'tabah and encompassing sections of the Al-Dhala, Qa'tabah, and Al-Husha districts. The second is located south of Al-Dhala city and stretches southwest along the Wadi Tabagyan catchment region in Al-Azerq district, which is a tributary of Wadi Tuban.

The Al-Dhala Qatabah Basin's morphology ranges from flat plains to steep slopes and hills made up primarily of restricted volcanic rocks; sands and outwash sediments blanketed the wadi basin. The Al-Dhala Qatabah Basin is located between 1100 and 1800 meters above sea level and receives about 269 mm of rainfall per year, as well as significant recharge from nearby mountain drainage.

The delineation of fluoride contamination areas from Taiz governorate has been conducted, and the iso-line contour map of fluoride ion concentration from At Aaiziyah district and its surrounding villages is given in **Figure 2**.

It's clearly observed that the villages of Jabal Sabir, Hawban, Hethran, and Al-Bryehey as well as Taiz City are the most affected areas by fluoride contamination in groundwater.

Some of Sana'a governorate districts, particularly Sanhan, had the highest fluoride concentrations in their drinking water (UNICEF, 2008). The majority of Yemenis living in rural areas rely on deep well water for drinking and cooking, and many of these wells are contaminated with fluoride in concentrations ranging from 2.5 to 32 mg/l. Fluorosis, particularly skeletal fluorosis, has never been seen in Yemen before, only about 8–10 years ago since it was first reported. Clinically, it develops as a result of

DOI: http://dx.doi.org/10.5772/intechopen.104255

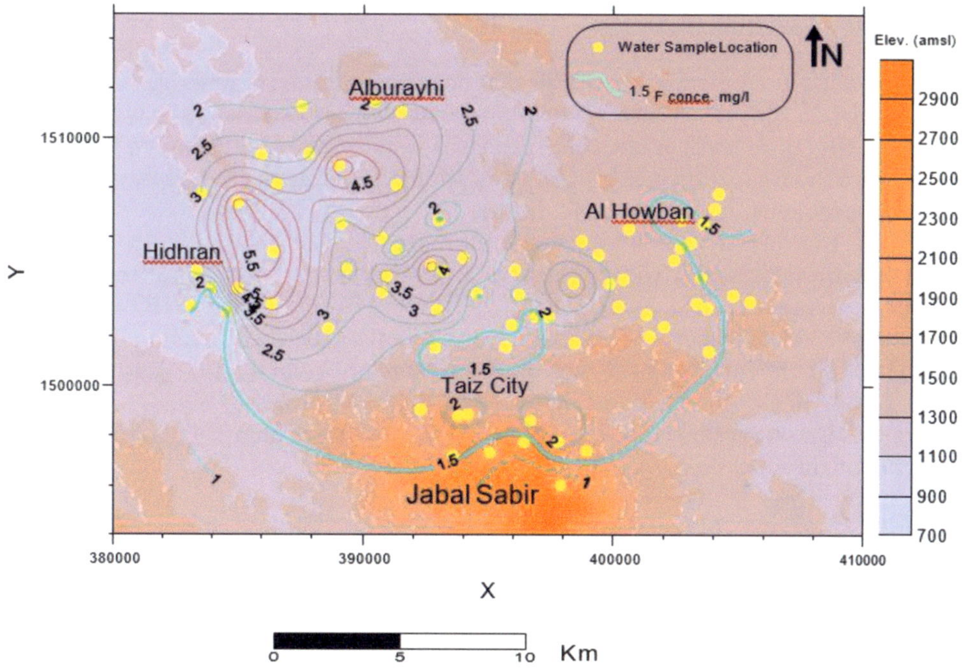

Figure 2.
Iso-line contour map of fluoride ion concentration from Taiz districts.

the high fluoride concentration in bones. Dental fluorosis, on the other hand, is not a new phenomenon in Yemen, particularly in the Taiz governorate [7].

A regional hydrogeochemical study from different Yemeni terrains indicated that water-rock interaction was most likely the primary cause of high ion concentrations in groundwater. According to geochemical modeling, the main minerals controlling the aqueous geochemistry of elevated fluoride ion contamination are calcite and fluorite. The concentration of F^- in groundwater was positively correlated with the concentrations of HCO_3^- and Na^+, indicating that groundwater with high concentrations of HCO_3^- and Na^+ leads to the dissolution of some fluoride-rich minerals. This situation of fluoride solubility control at higher fluoride concentrations can be explained by the fact that fluoride ions in groundwater can be increased as a result of $CaCO_3$ precipitation at high pH, which removes Ca^{2+} from solution and allows more fluorite to dissolve [20].

The groundwater data from different Yemeni areas reveals the chemical reactions that take place in the aquifer system. The results of the linear regression study on the relationship between F and HCO_3 (total alkalinity) from published papers show a positive connection, which could be owing to the simultaneous release of hydroxyl and bicarbonate ions during the leaching and dissolution of fluoride containing minerals into groundwater. With higher levels of alkalinity, the rate of weathering and mineral leaching rises, resulting in higher fluoride ion concentrations. High amounts of fluoride are also linked to greater Na^+ ion concentrations. This also favors that groundwater with high HCO_3^- and Na^+ content is usually alkaline and has relatively high OH^- content, so the OH- can replace the exchangeable F^- of fluoride-bearing minerals, increasing the F- content in groundwater [20].

5. Types of fluorosis in Yemeni regime

Fluoride is one of the few chemicals that have been proven to have harmful effects on people when consumed through drinking water. Fluoride in drinking water has beneficial effects on teeth at low concentrations, but excessive exposure to fluoride in drinking water, alone or in combination with fluoride from other sources, can cause a variety of problems. As the level and duration of exposure increase, these range from mild dental fluorosis to crippling skeletal fluorosis.

The chemical characteristics of the drinking water from the study reflect high fluoride contamination above the permissible limits of Yemeni, and WHO standers and most of them are poor whoever, their nutritional status is expected to be very poor. These factors including the high concentration of Na- and HCO_3^- and low concentration of Ca^{2+}ions will increase the severity of fluorosis.

The visual observations in some selected Yemeni villages identify that there are two types of fluorosis recognized, they are:

5.1 Dental fluorosis

Mottling of teeth is one of the most easily recognized symptoms from different Yemeni governorates especially Taiz and Al Dhalla. The teeth of the children in the affected areas lose their normal creamy white translucent color and become rough, opaque, and chalky white. Some of the local inhabitants indicated that their teeth have be extracted and replaced with dentures. The dental fluorosis ranges from mild to severe fluorosis. The photographs represent some of dental fluorosis from the different areas in Yemen. High percentage dental fluorosis among the children has been observed in Taiz and Al-Dhala basins (**Figure 3A** and **B**).

Dental fluorosis is the most common fluoride ailment identified in the afflicted areas, according to visual observations from selected villages. Fluorosis in the teeth can range from mild to severe.

In general, there is a link between fluoride in the water and the occurrence of dental fluorosis in the Taiz and Al-Dhala regions. A published research paper concluded that there is a positive relationship between fluoride in water and the occurrence of dental fluorosis in Sanhan, Taiz, and Al-Dhala regions [7].

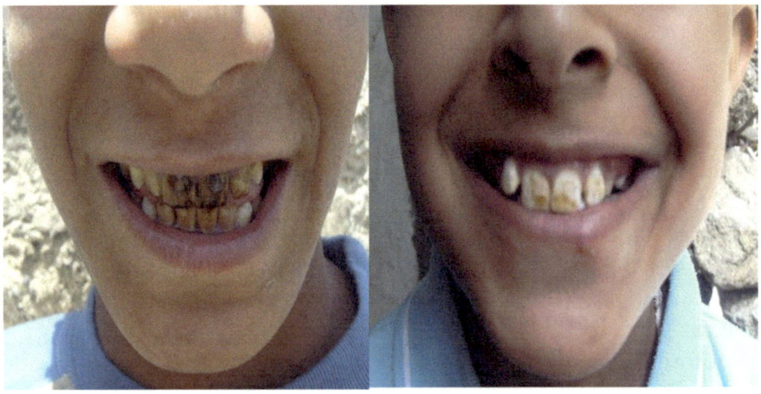

(a) (b)

Figure 3.
Figures present severe dental fluorosis (A) and moderate dental fluorosis (B) in Yemeni children in the affected area, after Alamry (2009) [7].

Figure 4.
A group of children showing skeletal deformation from Al-Dhalla region Yemen.

5.2 Skeletal fluorosis

Fluorapatite is 1000 times less soluble than hydroxyapatite, the most common mineral found in bone. The F^- ion aggressively substitutes for the OH^- ion, resulting in an accumulation of F^- in bone tissue and skeletal fluorosis [20].

The patient often complains of a vague discomfort in the limbs and trunk early on in the development of fluorotic changes in the skeleton. Back pain and stiffness, particularly in the lumbar region, follow. Fluorosis in the teeth is frequently visible with the naked eye and is easy to detect even by laypeople. On the other hand, even with the assistance of appropriate equipment, skeletal fluorosis and nonskeletal fluorosis are difficult to diagnose. Radiography is frequently used to detect symptoms such as joint enlargements or minor bone deformations.

There are no any publications presenting the problem of the skeletal fluorosis among children in the affected Yemeni places, while I have seen children in Al-Dhalla villages showing skeletal deformation, and these cases among children could be considered as cases of skeletal fluorosis unless proved otherwise, as it is shown in **Figure 4**.

6. Dietary practices of the children and fluorosis

The hydrogeochemical characteristics of the groundwater in the study area indicated that the volcanic and plutonic rocks are the primary sources of fluoride whoever, the food could be the secondary sources.

Fluoride intake during infancy and early childhood is mirrored in dental fluorosis patterns. In most cases, the fluoride content of drinking water is considered sufficient for determining the level of fluoride exposure in a given area. There has been evidence that fluoride uptake from other sources such as food, dust, and beverages is many times higher than that from water [21].

The percentage of children with fluorosis was found to be extremely high. Although high fluoride levels in drinking water may be to blame, various food habits (such as drinking black tea and chewing Gatof Catha edulis Forsk leaves (Khat)) indicated a high fluoride contribution to the diet. Some of the children used to chew Khat also, and the Khat is cultivated in the volcanic soil and irrigated by the high fluoride concentration water [22].

Cooking with fluoridated water raises fluoride levels significantly, particularly in dry foods such as maize flour, which absorbs a lot of water during cooking. It has been reported that the simultaneous intake of food and fluoride-containing compounds can affect fluoride availability in a positive or negative way, depending on the food type, mode of administration, and type of fluoride compound [21].

The diet consumed by the children was not balanced and lacked quality. It is composed of maize flour with milk and a few rare vegetables. Intake of milk and milk products is said to diminish the fluoride availability by 20–50% in humans. Although the area under study had children taking whole milk (boiled or fermented).

7. Conclusion

A high fluoride concentration has been reported in groundwater in tertiary volcanics, from Yemeni terrain. The major ion chemistry data from groundwater of this rock revealed that Na^+ is the most predominant cationic constituent followed by Ca^{2+} and Mg^{2+}, the HCO_3^- and SO^{4-} are found to be the most predominant anions followed by Cl^- and NO_3. High fluoride ion concentration in the Yemeni groundwater appears to be caused by high alkalinity due to HCO_3^- ions. Na^+ has a positive correlation with F^-, whereas Ca^{2+} has a negative correlation, resulting in an equilibrium condition in groundwater. $CaCO_3$ precipitation at high pH can increase fluoride ions in groundwater by removing Ca^{2+} from solution and allowing more fluorite to dissolve.

Dental fluorosis is the widely fluoride disease observed in the affected areas, whereas skeletal fluorosis is observed in some villages from Al-Dhalla region. The principal causes of fluorosis among the Yemeni population appear to be related to high fluoride concentration in drinking water and Khat chewing habits, which are cultivated in the volcanic soil and irrigated by the high fluoride concentration water.

Despite the high dental fluorosis prevalence in the affected areas, no restorative treatment is being carried out. Therefore, it is highly recommended to provide safe drinking water to fluoride-affected areas.

Acknowledgements

I would like to thank the National Water Resources Authority (NWRA) Yemen and UNDP for funding this research. Also I would like to thank the publisher and anonymous reviewers for valuable suggestions and comments that have enhanced the chapter quality.

Author details

Abdulmohsen Alamry
Faculty of Oil and Minerals, Department of Engineering Geology, Shabwah University, Ataq, Yemen

*Address all correspondence to: alamrygeo2@gmail.com

References

[1] Davis S, De Wiest R. Hydrogeology. Malabar, Fla.: Krieger Pub. Co.; 1991

[2] Rani J. Water quality analysis of HSIIDC industrial area Kundli, Sonipat Haryana. RA. Journal of Applied Research. 2016;**2**(10):678-684

[3] Elango L, Kannan R. Chapter 11 rock–water interaction and its control on chemical composition of groundwater. In: Developments in Environmental Science. Oxford: Elsevier; 2007. pp. 229-243

[4] García M, Borgnino L. CHAPTER 1. Fluoride in the context of the environment. Food and nutritional components in focus. In: Fluorine: Chemistry, Analysis, Function and Effects. Cambridge, UK: Royal Society of Chemistry; 2015. pp. 3-21

[5] Bozdağ A. Assessment of the hydrogeochemical characteristics of groundwater in two aquifer systems in Çumra plain, Central Anatolia. Environmental Earth Sciences. 2016;**75**(8):674

[6] Tóth J. Groundwater as a geologic agent: An overview of the causes, processes, and manifestations. Hydrogeology Journal. 1999;**7**(1):1-14

[7] Al-Amry A. Hydrogeochemistry and origin of fluoride in groundwater of Hidhran & Alburayhi Basin, northwest Taiz City, Yemen. Delta Journal of Science. 2009;**33**(1):10-20

[8] Shamsudduha M, Chandler R, Taylor R, Ahmed K. Recent trends in groundwater levels in a highly seasonal hydrological system: The Ganges-Brahmaputra-Meghna Delta. Hydrology and Earth System Sciences. 2009;**13**(12):2373-2385

[9] Kushawaha J, Aithani D. Geogenic pollutants in groundwater and their removal techniques. In: Groundwater Geochemistry: Pollution and Remediation Methods. New York: John Wiley & Sons Ltd; 2021. pp. 1-21

[10] Khichariya J, Verma Y. Geochemistry of fluoride enrichment in groundwater: A critical review of the Indian regime. International Journal of Chemical Engineering Research. 2021;**8**(1):1-4

[11] Raju N. Prevalence of fluorosis in the fluoride enriched groundwater in semi-arid parts of eastern India: Geochemistry and health implications. Quaternary International. 2017;**443**:265-278

[12] Satyanarayanan M, Balaram V, Al Hussin M, Al Jemaili M, Rao T, Mathur R, et al. Assessment of groundwater quality in a structurally deformed granitic terrain in Hyderabad, India. Environmental Monitoring and Assessment. 2007;**131**(1-3):117-127

[13] Aqeel A, Al-Amry A, Alharbi O. Assessment and geospatial distribution mapping of fluoride concentrations in the groundwater of Al-Howban Basin, Taiz-Yemen. Arabian Journal of Geosciences. 2017;**10**(14):312

[14] Nas B, Berktay A. Groundwater quality mapping in urban groundwater using GIS. Environmental Monitoring and Assessment. 2008;**160**(1-4):215-227

[15] Gibbs R. Mechanisms controlling world water chemistry. Science. 1970;**170**(3962):1088-1090

[16] Reddy D, Nagabhushanam P, Sukhija B, Reddy A, Smedley P. Fluoride dynamics in the granitic aquifer of the Wailapally watershed, Nalgonda

District, India. Chemical Geology. 2010;**269**(3-4):278-289

[17] Jha S, Mishra V, Sharma D, Damodaran T. Fluoride in the environment and its metabolism in humans. Reviews of Environmental Contamination and Toxicology. 2011;**211**:121-142

[18] Saeed Zango M, Pelig-Ba K, Anim-Gyampo M, Gibrilla A, Abu M. Assessment of the mineralogy of granitoids and associated granitic gneisses responsible for groundwater fluoride mobilization in the Vea catchment, upper east region, Ghana. Sustainable Water Resources Management. 2021;**8**(1):213

[19] Cao H, Xie X, Wang Y, Liu H. Predicting geogenic groundwater fluoride contamination throughout China. Journal of Environmental Sciences. 2022;**115**:140-148

[20] Al-Amry A. Geochemical process controlling the elevated fluoride concentrations in groundwater of Al-Azareq Basin, Al-Dhala, Yemen. University of Aden, Journal of Natural and Applied Sciences. 2011;**15**(1):111-120

[21] Trautner K, Einwag J. Influence of milk and food on fluoride bioavailability from NaF and Na2FPO 3 in man. Journal of Dental Research. 1989;**68**(1):72-77

[22] Al-Amry AS, Habtoor A, Qatan A. Hydrogeochemical characterization and environmental impact of fluoride contamination in groundwater from Al-Dhala Basin, Yemen. Electronic Journal of University of Aden for Basic and Applied Sciences. 2020;**1**(1):30-38. Available from: https://ejua.net/index.php/EJUA-BA/article/view/8

Chapter 2

Fluoride Geochemistry and Health Hazards: A Case Study

Babu Rao Gudipudi

Abstract

This chapter was aimed to identify the relationship between fluoride (F) enrichment and prevalence of endemic fluorosis in a rural area of Nuzendla mandal in Guntur District, Andhra Pradesh, India. The concentration of F varies from 0.5 to 12.4 mg/L in pre-monsoon groundwater and 0.14 to 16.0 mg/L in post-monsoon groundwater in the collected and analyzed fifty water samples. Dental survey conducted in the study area based on Dean Classification Index indicated different degrees of dental fluorosis due to the varying concentrations of F in drinking water. The significant positive correlation is identified between the F content of groundwater and urine fluorosis-affected children. The F level in urine suggests that a high level of endemic fluorosis is prevalent in the Nuzendla mandal due to the consequence of a higher concentration of F in underground aquifers. This study concludes that the high concentration of F in groundwater leads to increased dental deformities among the surveyed people and also urinary F is a good indicator of community exposure F.

Keywords: fluoride, Nuzendla mandal, dean classification index, fluorosis, community fluorosis Index

1. Introduction

Fluoride (F) occurs in rocks, soil, air, water, plants, and animals as well as in human body. Fluoride content in subsurface water is controlled by temperature, pH and solubility of F-bearing minerals. The subsurface water, most of which originates from rainfall or surface water bodies, gains minerals during its transport and residency period of earth crust [1, 2]. Continuous intake of F contaminated groundwater (>1.5 mg/L) without proper treatment cause chronic endemic fluorosis. There is a close relationship between environmental F and general health. Fluoride deficiency increases incidences of dental caries among the population, while the excess F intake causes dental, skeletal fluorosis, and other forms of non-skeletal tissue fluorosis. Hydrofluorosis is a major toxicological and public health problem in water-stressed regions.

Fluorosis continues to be an endemic problem around the world. Moderate levels of F ingestion reduce incidences of dental caries and also promote healthy development of bones and teeth [3, 4]. Hydroxylapatite is main mineral phase of the human teeth enamel. Dental fluorosis, which is characterized by mottling of tooth surface, is the most adverse effect of overexposure to dietary F. Fluoride accumulates in dentin, which is the mineralized tissue underlying the enamel, and its chronic overexposure

IntechOpen

could cause dentin to crack more easily [4]. Children within the age group of 0–12 years are most prone to fluorosis as their body tissues are in formative stage.

Fluorosis, which was initially considered to be a problem of teeth only, has now turned to be a serious health hazard affecting many other body systems manifesting through joint pains, muscular pain; skeletal deformations, and malformations characterized by increased in bone mass and density, pain and stiffness in backbone, hip region, and other joints. This is because continuous intake of high F causes ligaments of spine become calcified and ossified [5]. Studies indicate that F intake could increase probability of cancer in the kidney and bladder based on tendency for hydrogen fluoride (HF) to form under the acidic conditions such as urine [6].

Fluoride occurs in natural waters mainly in the form of F, whose concentrations may range up to 2800 mg/L [7]. Fluoride levels are high in groundwater where the source minerals such as amphiboles, micas, fluorapatite, topaz, cryolite, certain clays, and villiaumite [8]. Enrichment of F in Geological substrate is from the fluorite mineral phase in the rocks along with the weathering of rocks.

Hydrochemical techniques are normally employed in the water quality management. In these techniques, the data regarding the origin and behavior of major cations (Ca^{2+}, Mg^{2+}, Na^{+}, K^{+}) and anions CO_3^{2-}, HCO_3^{-}, Cl^{-}, SO_4^{2-}) in the groundwater permits the elucidation of the hydrogeochemical compositions of the water [9]. This generally varies depending upon the solubility of the chemical components from the dissolution of the mineral component of the rocks that host the aquifer.

The current work was aimed to investigate the relationship between the consumption of water from natural high F terrain and the prevalence of dental fluorosis in the study area.

2. The study area

2.1 Location and climate of the study area

Nuzendla mandal is the present study area, which is located in Narasaraopet Revenue Division of Guntur District, Andhra Pradesh, India (**Figure 1**). It lies in

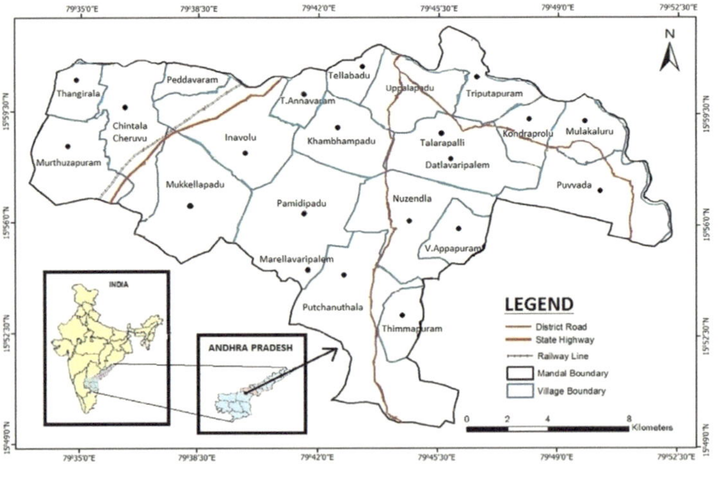

Figure 1.
Location map of Nuzendla mandal of Guntur district, Andhra Pradesh, India.

between the latitudes 79°33′28″–79°52′51″ E, and longitudes 15°49′26″ and 16°01′42″ N and extends to an area of about 350 km^2 and is distributed in 20 rural villages. The area experiences a semi-arid climate, with minimum and maximum temperatures of 16.8°C and 48.5°C, respectively. Rentachintala of Guntur district (nearest IMD station) records the highest temperatures (48.5°C) during summer (March to May). The daily sunshine hours range from 3.5 to 10.5, with a mean of 7.5. The relative humidity is from 30% to 80% with a mean of 52%. The mean wind velocity ranges from 4.7 to 16.3 km/h, with an average of 10.5 km/h. The wind velocity is higher during the southwest monsoon compared to the rest of the period. The average annual rainfall for a period of 12 years (1991–2013) is 718.38 mm. The semi-arid climate of the study area with average annual rainfall initiates the evaporation process which plays a crucial role in the release of F fluoride from underlying rocks into the groundwater.

3. Geology of the study area

The study area is under laid by the rock formations ranging from Archaean to Permo-carboniferous age (**Table 1**). The rock formations include quartz-mica-schist, banded-biotite-hornblende-gneiss/granite, and coarse-grained sandstone (**Figure 2**). Major part of the study area is occupied by quartz-mica-schist. The second dominant rocks in the study area are biotite-hornblende granitic/gneisses with migmatite and pegmatite patches, which are observed from the northern part of the study area. A small portion in the eastern and southern parts is covered by coarse-grained sandstones [11]. A very small area in the eastern part is occupied by gabbro. Mineralogically, quartz-mica-schist is composed principally of quartz and mica, and generally of muscovite and biotite. The modal composition of quartz-mica schist shows 25% of muscovite, 15% biotite, 30% orthoclase, 20% quartz, and 10% plagioclase and chlorite, together with some opaque minerals [12]. The biotite-hornblende granitic/gneisses appear in their light and dark color banded texture. The light color band is composed of quartz and plagioclase feldspar, while the dark color is composed of biotite, hornblende, and opaque's. The modal composition of biotite-hornblende granitic/gneisses have 34.25% of quartz, >3.75% k-feldspar, 20.35% of plagioclase, 35% of biotite,

Super group	Group	Lithology	Age
	Gondwana	Coarse-grained Sandstone	Permo-carboniferous
Unconformity			
—	Acid/mafic intrusives	Gabbro/Norite Alkali feldspar granite	Paleo-Proterozoic to Meso-Proterozoic
Eastern Dharwar Carton	Udaigiri group	Quartz-biotite-Muscovite-chlorite schist	Archaean to
	Peninsular gnessic complex	Banded-biotite- hornblende-granite/gneiss-with migmatite patches	Paleo-Proterozoic

Table 1.
Geological succession of the study area [10].

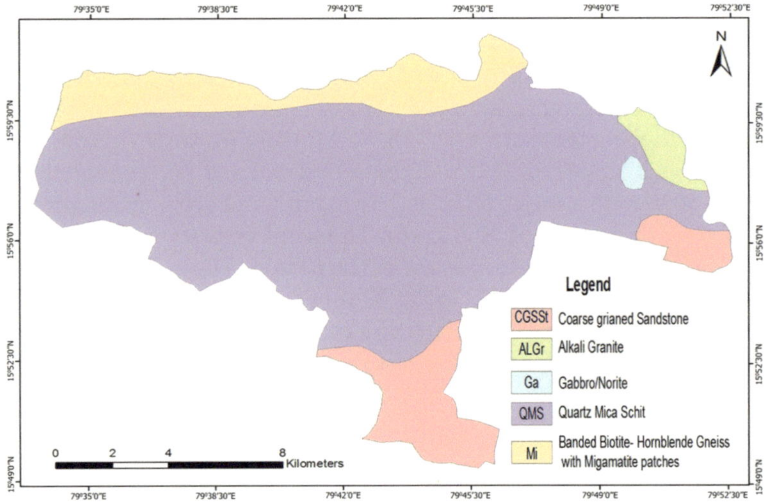

Figure 2.
Geological Map of Study area.

5.2% hornblende, 0.5% chlorite,0.15% of sphene, 0.15% of zircon, 0.3% of epidote and 0.10% fluorite and 0.4% opaques [13]. Since the coarse-grained sandstones of Gondwana age are considered to be the weathered product of Eastern Dharwar Craton, it is essentially composed of quartz and little proportion of feldspars, together with accessories of biotite, hornblende, apatite, epidote, fluorite, sphene, zircon, etc. The volume percentage of the different minerals depends on the cementing material and environment of deposition. The muscovite and biotite micas, fluorite minerals contribute to the higher F levels in this study.

4. Hydrogeology of the study area

Generally, groundwater occurs in all the formations of the study area. But it occurs under phreatic conditions in the weathered and fractured rocks at shallow depths and under semi-confined to confined conditions in the deeper fractured rocks. Development of the aquifer conditions in the quartz-mica schist and banded-biotite-hornblende-gneiss is generally less due to lack of primary porosity. However, the occurrence and movement of groundwater in the rocks depend on the development of extent of weathered rock portions and degree of the fractures in the rocks. The depth of weathering in the rocks is from 2 to 12 m below ground level (bgl) and the fractured rocks from 3 to 32 m bgl. Development of groundwater is through shallow wells (dug wells) and deep wells (bore wells/tube wells) in the study area. The depth of dug wells varies from 5.50 to 18.50 m bgl.

5. Materials and methods

5.1 Methodology

The methodology comprises of field procedures and analytical techniques. The field procedures include mapping techniques, well inventory and collection of groundwater samples and survey on health implications caused by F. The analytical techniques include the determination various physicochemical parameters of collected groundwater samples.

5.2 Field procedures

The mapping techniques covered geological mapping, demarcation of geomorphological features and preparation of slope, soil, drainage and land use/land cover maps. This work was carried out, using the Survey of India toposheets of 56P/12, 56P/16, 57M/9, and 57M/13 on a scale of check 1: 50,000. Indian Remote Sensing Satellite (IRS) ID Linear Imaging Self Scanner (LISS) III of geocoded false color composite of December 9, 2012 on 1:50,000 scales are used to get information on soils, geomorphological features, lithology, lineaments, and land use/land cover with a limited ground truth. Geological mapping was carried out by marking the contacts between the geological formations as well as the structural features. Geomorphological features were demarcated based on the field observations and the available literature.

Fifty groundwater samples were collected in pre-monsoon (month of May) and post-monsoon (November) seasons during the year 2012 in the study area. Prior to water sampling, sampling bottles soaked in 1:1 HCl for 24 h were rinsed with distilled water, followed by deionized water. They were washed again prior to each sampling of the filtrates. The bottles were tightly capped to protect the samples from atmospheric CO_2, adequately labeled, and preserved in the refrigerator till they were taken to laboratory for measurement. Data on location of wells, geographic coordinates, type of well, depth to groundwater level, and water taste was collected. The variations in the groundwater levels in the wells were recorded, using a water level recorder.

The people living in the study area suffering from different stages of fluorosis by consuming fluoridated water were identified. The dental fluorosis stages were identified by adopting Dean's classification [10]. The fluoride levels are examined in the human body through the analysis of urine samples of the effected persons.

5.3 Analytical techniques

The collected groundwater samples from the field were analyzed for chemical variables, using the standard water quality methodology of the American Public Health Association [10]. The chemical variables include pH, electric conductivity (EC), total dissolved solids (TDS), calcium (Ca^{2+}), magnesium (Mg^{2+}), sodium (Na^+), potassium (K^+), bicarbonate (HCO_3^-), chloride (Cl^-), sulphate (SO_4^{2-}), nitrate (NO_3^-) and fluoride (F^-). The pH and EC of the groundwater samples were measured in the field, using a portable pH meter (60510-ISE, YSI Pro plus) and EC meters (60530-ISE, YSI Pro plus). The TDS was calculated from EC adhering to the procedure of conversion factor adopted by Hem [14]. The rest of the chemical variables were determined in the laboratory immediately after the groundwater sampling. A summary of the analytical procedures is listed in **Table 2**. All concentrations of chemical parameters are expressed in milligrams per liter (mg/L), except pH (units) and EC (μS/cm at 25°C).

5.4 Hydrogeochemical facies

The concept of hydrogeochemical facies has been used here to provide a model for explaining the distribution and genesis of principal types of groundwater, as it

Chemical parameters	Methods
Bicarbonate (HCO_3^-)	Titration with HCl
Calcium (Ca^{2+})	Titration with EDTA
Carbonate (CO_3^{2-})	Titration with HCl
Chloride (Cl^-)	Titration with $AgNO_3$
Fluoride (F^-)	Spectrophotometer
Hydrogen ion concentration (pH)	pH meter (60510-ISE, YSI Pro plus)
Magnesium (Mg^{2+})	Calculation (TH- Ca^{2+})
Nitrate (NO_3^-)	Colorimeter
Potassium (K^+)	Flame photometer
Silica (Si)	Spectrophotometer
Sodium (Na^+)	Flame photometer
Specific Electrical Conductivity (SEC)*	SEC meter (60530-ISE, YSI Pro plus)
Sulphate (SO_4^{2-})	Spectrophotometer
Total dissolved solids (TDS)	SEC X Conversion factor (0.64)

Table 2.
Methods used for chemical analysis of groundwater.

reflects the response of chemical processes in a lithological framework and the pattern of water flow in it [15, 16].

5.5 Piper's trilinear diagram

A Piper trilinear diagram was used in understanding the hydrogeochemical characteristics of groundwater in the area [17]. It consists of two triangles, one for plotting cations and the other for plotting anions, and one diamond-shaped field from which hydrochemical facies were identified.

5.6 Dental survey

A questionnaire pre-format prescribed by Rajiv Gandhi Drinking Water Mission [18] and earlier described by Dahyia et al. [19] was used to score the incidence and degree of manifestation of dental fluorosis. Clinical dental examination was executed rendering to the requirements defined by the World Health Organization Formational Oral Health Surveys [20] by taking 10 minutes as an orientation period spell for the basic examination of a child. The test area was prepared with the required hygiene and safety measures, using previously sterilized instruments and having easy access to sterilization procedures, and using a plane mirror and a periodontal probe. Community fluorosis index (CFI) was calculated based on equation [1] as

$$CFI = \frac{\text{Number of people} \times \text{Deans numerical weight}}{\text{Total number of people}} \tag{1}$$

The symptoms of dental fluorosis among the communities were recorded using, randomized sampling method. The results were classified into seven categories based on the Dean's classification viz., normal, questionable, very mild, mild, moderate, moderately severe, and severe. The classifications were given a numerical weights of 0.0, 0.5, 1.0, 1.5, 2.0, 3.0, and 4.0, respectively, in order of increasing severity [21–24].

5.7 Urine sample collection and analysis

A total of 50 urine samples (one sample from each location where the groundwater samples were collected) were collected from the children of same age group (10–12 years age group). The samples were further classified into high (>1.5 mg/L), intermediate (0.6–1.5 mg/L), and low F (<0.6 mg/L) based on groundwater F content. Pre-labeled 500-ml plastic-capped disposable bottles (prewashed and dried containing 0.2 g of ethylene diamine tetra-acetic acid, EDTA) were distributed to the selected persons in the villages of the study area and brought to the laboratory in an ice box and stored at 4°C in a refrigerator. EDTA (0.2 g) was added to check and minimize the interference from complexation of F by cations such as calcium. The samples were analyzed for F content using the 2-(parasulfophenylazo)-1,8-dihydroxy-3,6-naphthalene-disulfonate SPADNS method. The individuals were also explained the importance of the program and were motivated to cooperate in this study. An informed consent was obtained from the participants. Information on the drinking water sources, dietary practice, period of living in a particular location, and other related data were collected through an open-ended questionnaire.

6. Results and discussion

6.1 Hydrogeochemical evolution

A trilinear diagram is widely used in understanding the hydrogeochemical evolution of groundwater [17]. The diagram consists of two triangles and one diamond-shaped field. The left side triangle is for plotting of cations (Ca^{2+}, Mg^{2+} and Na^{+} + K^{+}) and the right side triangle for plotting of anions (HCO_3^{-} + CO_3^{2-}, Cl^{-}, and SO_4^{2-}) expressed in

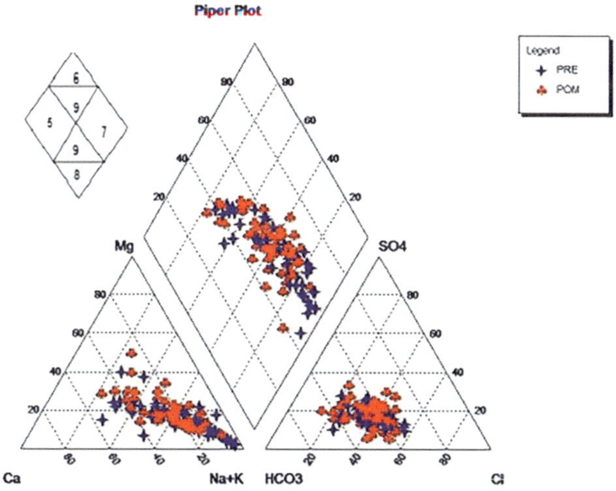

Figure 3.
Hydrogeochemical facies during pre-and post-monsoon periods.

F– Range (mg/L)	Pre-monsoon	MeanCa^{2+} (mg/L)	MeanNa^{+} (mg/L)	Mean HCO_3^{2-} (mg/L)	Mean TDS (mg/L)	% of samples	Post-monsoon	Mean Ca^{2+} (mg/L)	Mean Na^{+} (mg/L)	Mean HCO_3^{2-} (mg/L)	Mean TDS (mg/L)	% of samples
< 0.6	Ca > Mg > Na > K: HCO_3 > NO_3 > Cl > SO_4	156	87	308	981	2	Ca > Mg > Na > K: HCO_3 > NO_3 > Cl > SO_4	168	132	337	1192	6
0.6 – 1.5	Na > Ca > Mg > K: HCO_3 > Cl > NO_3 > SO_4	117	216	346	1,207	38	Ca > Na > Mg > K: HCO_3 > Cl > NO_3 > SO_4	156	182	375	1265	28
>1.5	Na > Ca > Mg > K: HCO_3 > Cl > NO_3 > SO_4	83	338	427	1,445	60	Na > Ca > Mg > K: HCO_3 > Cl > NO_3 > SO_4	66	456	461	1655	66

Table 3.
Hydrogeochemical facies of groundwater during the pre-and post-monsoon periods of study area.

percentage. The diamond-shaped field (consisting of the total cations and anions), which is the upper side of these two triangles is used for representing the overall chemical quality of groundwater. The Zone-5 represents carbonate hardness (Ca^{2+}: HCO_3^- type), zone 6 non-carbonate hardness (Ca^{2+}: Cl^- type), the zone-7 non-carbonate alkali (Na^+: Cl^- type), the zone-8 carbonate alkali (Na^+: HCO_3^- type), and the zone 9 mixed types. The chemical data of the groundwater samples are plotted in the Piper's diagram (**Figure 3**). Most groundwater samples fall in the center as well as in the right lower corner of the cation triangle in both the seasons. It indicates the high concentration of Na^+ in the groundwater.

Most of the anions in pre-and post-monsoon groundwater samples fell in center of the triangle representing HCO_3^- type. Therefore, the groundwater is dominated by Na^+-HCO_3^- facies in general, which is further supported by hydrogechemical facies (**Table 3**). In the centrally located diamond-shaped field, the groundwater samples fall in zones 5–9. It suggests that the fresh water (zone 5) moves towards saline water (zone 7) through the zones of 6–8, following the flow path. That means the initial water quality is controlled by water-rock interaction and is subsequently modified by anthropogenic sources. Because of this, the concentrations of Na^+ and Cl^- increase, which enhance the TDS content, are including the F content in the groundwater.

6.2 Mechanisms controlling groundwater chemistry

To understand the groundwater interaction with precipitation (rainfall), rock, and evaporation as mechanisms controlling the water chemistry [25], the ratios for major cations ($Na^+ + K^+$: $Na^+ + K^+ + Ca^{2+}$) and for major anions (Cl^-: $Cl^- + HCO_3^-$) computed from the ionic concentration of groundwater of the study area are plotted against TDS (**Figure 4**).

Most groundwater samples fall in the rock domain in both the seasons, where the TDS is between 100 and 1000 mg/L (**Figure 4**). The remaining groundwater samples

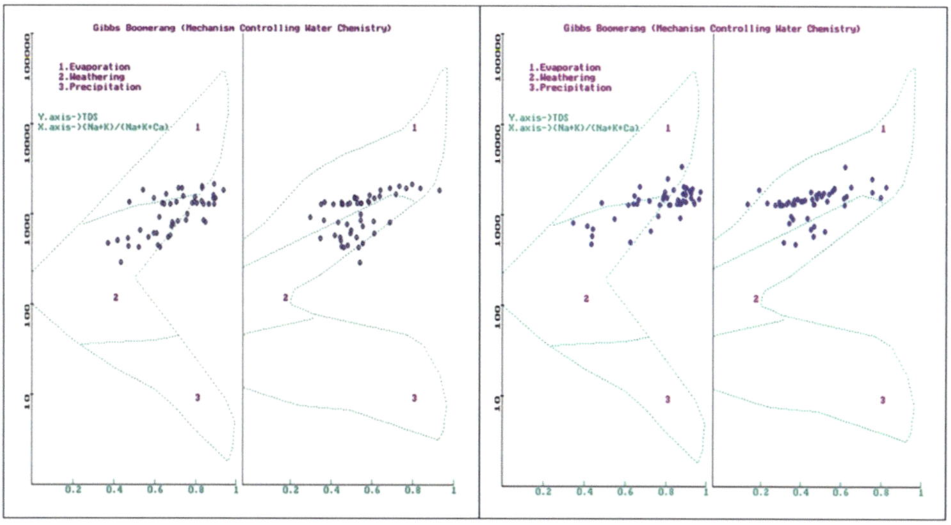

Figure 4.
Mechanisms controlling groundwater chemistry (after Gibbs [25]).

are observed from the evaporation domain, where the TDS is more than 1000 mg/L. Falling off the groundwater samples in the rock domain indicates the water-rock interaction. The average values of TDS, Na^+, HCO_3^-, and Cl^- vary from 844 to 981, 107.8 to 94.5, 271 to 284.75, and 95.4 to 88.5 mg/L from pre- to post-monsoon, where the TDS is less than 1000 mg/L, while they are from 1402 to 1583, 306.68 to 380.54, 408.2 to 430.17, and 239.2 to 278.85 mg/L in the respective seasons, where the TDS is more than 1000 mg/L (**Table 4**). The increase of Na^+ and Cl^- from TDS less than 1000 mg/L to TDS more than 1000 mg/L concentrations are mainly caused by anthropogenic pollution. Because of this reason, the groundwater samples move towards the evaporation domain from the rock domain, as also reported by Wang et al. [26], Mamatha and Rao [27], Li et al. [28], and Narasimha and Sudarshan [29] in other regions (**Figure 4**).

Since the groundwater quality is dominated by Na^+ and HCO_3^- ions due to rock-water interaction, this factor appears as governing process for the release of F from the country rocks. As a result, the groundwater shows the higher F content. Similar conditions have been reported by Li et al. [30] in China. On the other hand, the evaporation and/or anthropogenic activity increases the Na^+ and Cl^- contents, which make the higher TDS.

6.3 Human health survey

Human health survey has been conducted in the present study area to analyze the fluorosis hazards with respect to F^- content in the selected endemic villages of T. Annavaram, Talrapalli, Datlavaripalem, Marellavaripalem, and Upplapadu. Dental health survey collected 659 data samples on people, including the males (213), females (214), and children (232) to understand the severity of fluorosis hazard in this area. The results of dental survey carried out in the selected villages of study area are presented in **Table 5**.

The results of dental survey population are compared with the Dean's Classification Index (1942) of tooth surface (15; **Table 6**). The results of Dental Fluorosis Index (DCI) and Community Fluorosis Index (CFI) are presented in **Table 7**. Out of the 76 members surveyed in T. Annavaram, the people who come under questionable, very mild, mild, moderate, moderately severe, and severe categories are 2, 7, 12, 16, 11, and 6 respectively. These contribute 71.05% of the fluorosis (**Figure 5a**). In Talrapalli, 180 people are surveyed. The mentioned categories are 20, 18, 14, 14, 4, and 4 respectively, which contribute 41.11% of the fluorosis (**Figure 5b**). Datlavaripalem records the highest dental hazard in the people in the respective categories are 17, 32, 19, 22, 17, and 21 (**Figure 5c**). These are contributes 88.27% of the fluorosis. In the Marellavaripalem, the total surveyed people are 155. Out of which, the questionable, very mild, mild, moderate, moderately severe, and severe categories are 15, 18, 14, 12, 11, and 14 respectively, contributing 74.33% of

TDS range (mg/L)	Na^+ (mg/L)		Cl^- (mg/L)		$HCO_3^- + CO_3^{2-}$ (mg/L)	
	Pre-monsoon	Post-monsoon	Pre-monsoon	Post-monsoon	Pre-monsoon	Post-monsoon
<1000	107.80	101.00	95.4	87.00	328.40	353.00
>1000	306.73	365.30	239.20	269.85	479.46	500.91

Table 4.
Classification of Na^+, Cl^- and HCO_3^{2-} + CO_3^{2-} based on TDS range.

Name of the village	Population surveyed	Adults						Children		
		Male			Female			Children		
		People surveyed	Effected people		People surveyed	Effected people		Surveyed Children	Effected Children	%
			People	%		People	%			
T.Annavaram	76	30	26	86.67	30	18	60	16	10	62.5
Talrapalli	180	54	22	40.74	54	11	20.37	72	41	56.94
Datlavaripalem	155	43	37	86.05	44	33	75	68	58	85.29
Marellavaripalem	113	41	28	68.29	41	32	78.05	31	24	77.42
Upplapadu	135	45	14	35	45	10	25	45	19	42.22
Total	659	213	127	—	214	104	—	232	152	—
Average				63.35			51.68			64.87

Table 5.
Results of dental survey in the selected villages of study area.

Deans number	Category	Indication of tooth surface
0	Normal	Translucent, smooth enamel with a glossy appearance
0.5	Questionable	Seen in endemic areas, borderline between normal and very mild
1	Very mild	Small opaque, paper-white areas scattered irregularly over the labial and buccal surface of teeth
1.5	mild	White opaque areas are more extensive but do involve many surfaces
2	Moderate	Entire tooth surface involved, minute pitting often present on labial and buccal surfaces, brown surface, brown stains, frequently disfiguring
3	Moderately severe	Entire tooth surface involved marked pitting with intense brown stain
4	Severe	Widespread, deep brown or black areas, corrosion type of mottled enamel

Table 6.
Dean's classification [14].

the dental fluorosis (**Figure 5d**). The lowest dental fluorosis (34.29%) is recorded in the Upplapadu village (**Figure 5e**) where the questionable, very mild, mild, moderate, moderately severe, and severe categories are 5, 8, 10, 6, 3, and 11 respectively. The above data indicate the different degrees of fluorosis according to varying concentrations of F in drinking water. In the present study area, the concentration of F varies from 0.5 to 12.4 mg/L in pre-monsoon groundwater and 0.14 to 16.0 mg/L in post-monsoon groundwater.

Figure 6 indicates that the dental fluorosis in the surveyed villages is high in children (64.87%) compared to men (63.35%) and women (58.38%). This could be due to effect of the drinking water on children, in particular as their body tissues are in their growth stage [31, 32]. The effect of fluorosis is higher in males compared to female. Generally, the males take more drinking water and diets than the females due to their greater physical activity. This is also supported by the significant positive correlation between average F content in groundwater and the percentage of dental fluorosis.

6.4 Community fluorosis index

Community Fluorosis Index (CFI) was calculated based on the symptoms of dental fluorosis with respect to DCI [14, 23]. Criteria for people with symptoms of dental fluorosis are identified and classified in each category based on CFI. CFI is the ratio of the number of people affected in each category and Dean's numerical weight to total number of affected people (Eq. (1)). If CFI is greater than 0.6; fluorosis is considered to be a public health problem in that area [14, 23–25].

Higher prevalence rates of endemic fluorosis are observed in four out of five screened villages. The CFI values of T. Annavaram, Datlavaripalem, Marellavaripalem, and Uppalapadu are 1.51, 1.60, 1.41, and 0.86 respectively which may cause public health problems. Tarlapalli village is the only one showing CFI value (0.58) is less than 0.6, where there is no fluorosis hazard was observed (**Table 7**).

The above dental survey in the study area indicates that there is different degree of fluorosis hazard due to varying concentration of F in drinking water, quantity of water consumption, intake of nutrients at risk, dietary substance, hot climate condition and long period exposure after digestion of in human body [33].

DOI: http://dx.doi.org/10.5772/intechopen.105156

Village	AverageF^- Value	Normal	Questionable	Very Mild	Mild	Moderate	Moderately severe	Severe	Total	CFI	% of DF
1	3.38	26	2 (1)*	7 (7)*	12 (18)*	16 (32)*	11 (33)*	6 (24)*	76	1.51	71.05
2	1.56	106	20 (10)	18 (18)	14 (21)	14 (28)	4 (12)	4 (16)	180	0.58	41.11
3	6.08	27	17 (8.5)	32 (32)	19 (28.5)	22 (44)	17 (51)	21 (84)	155	1.60	88.27
4	3.7	29	15 (7.5)	18 (18)	14 (21)	12 (24)	11 (33)	14 (48)	113	1.41	74.33
5	1.93	92	5 (2.5)	8 (8)	10 (15)	6 (12)	3 (9)	11 (44)	135	0.86	31.85

**Figures shown in the brackets indicates the Calculated DCI values for each individual category.*
Village names: 1; T. Annavaram; 2. Talrapalli; 3. Datlavaripalem; 4. Marellavaripalem; 5. Uppalapadu.

Table 7.
Results of dental fluorosis and community fluorosis index.

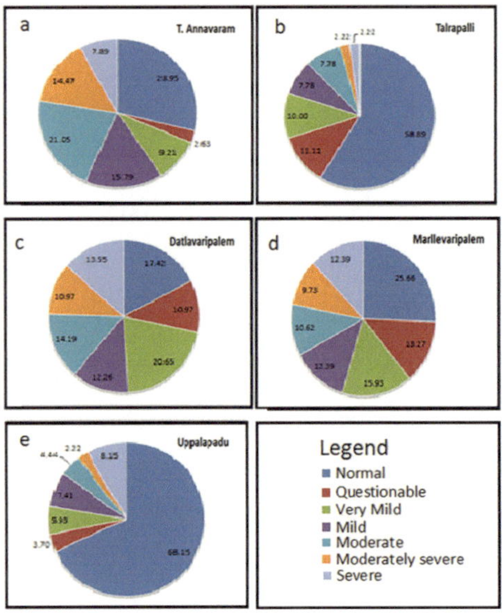

Figure 5.
Pie plot showing the different degrees of dental hazard in the study area.

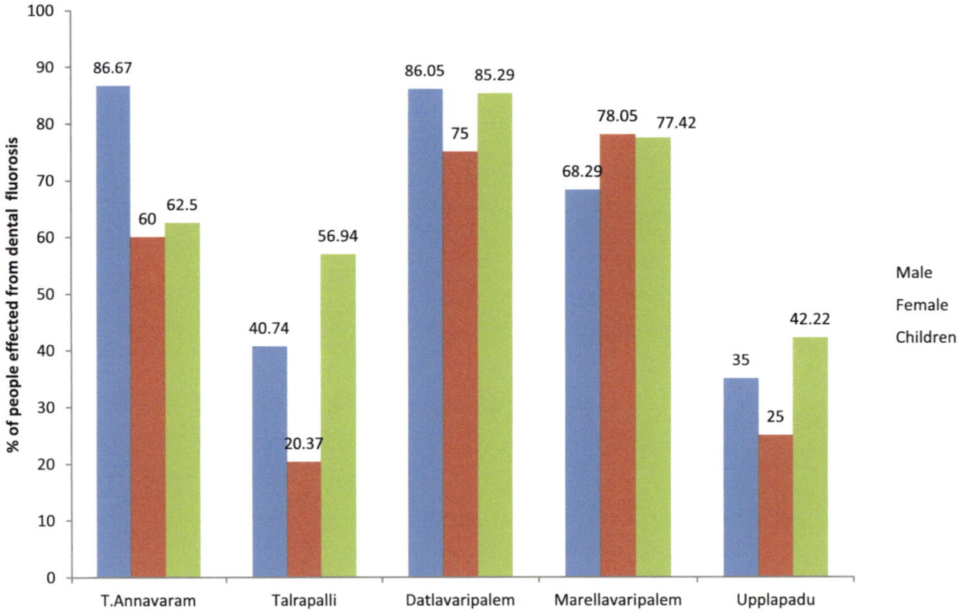

Figure 6.
Histogram showing the dental hazards in the study area.

6.5 Urine sample analysis survey

The urine sample analysis survey is carried out alongside, the dental survey. F is excreted primarily through urine [34], which is an early indicator of fluoride poisoning. The F content in urine depends on the concentration of F in the drinking water. The acceptable concentration of urine F is 1.0 mg/L [35]. The results of the urine samples of current study area showed that minimum and maximum urinary F concentrations are 1.2 and 16.2 mg/l, while the groundwater samples (mean of both seasons) show 0.42–14.2 mg/l, respectively (**Table 8**). The urine samples were further classified as low (<0.6 mg/L), intermediate (0.6–1.5 mg/L) and high (>1.5 mg/L) based on mean F− content in the groundwater.

The minimum and maximum of urine F concentrations among the low (<0.6 mg/L), intermediate (0.6–1.5 mg/L) and high (>1.5 mg/L) F areas ranged from 0.45 to 1.2 mg/l, 1.4 to 4.2 mg/L and 2.1 to 16.2 mg/L respectively. The corresponding mean values were 1.2 mg/L, 2.52 mg/L, and 4.93 mg/L respectively (**Figure 7**). The lowest urine concentration (1.2 mg/L) is observed in low F areas (with F < 0.6 mg/L in water) and the highest urine concentration (16.2 mg/L) was observed among areas of high F concentration (>1.5 mg/L) in water. This was also supported by excellent positive correlation between the urinary and groundwater F (**Figure 8**). The mean F concentration in urine has enhanced from low to high F groups (**Table 8**). The urine F content even in the low F areas exceeded the acceptable concentrations of 1.0 mg/L. This shows that the groundwater consumed by the individuals was the main causative factor for fluorosis hazard.

F^- range in groundwater	Range and mean F^- content in groundwater (mg/L)	Range and mean F^- content in urine (mg/L)	% of samples
<0.6 mg/L	0.45	1.2	4
	(0.14–0.50)	(1.2)	
0.6–1.5 mg/L	1.14	2.52	32
	(0.76–1.14)	(1.4–4.2)	
>1.5 mg/L	2.75	4.93	64
	(1.55–14.20)	(2.1–16.2)	

*Mean F^- concentration of groundwater for pre- and post- monsoon seasons.

Table 8.
Data on different F^- concentrations in groundwater and urine (mg/L).*

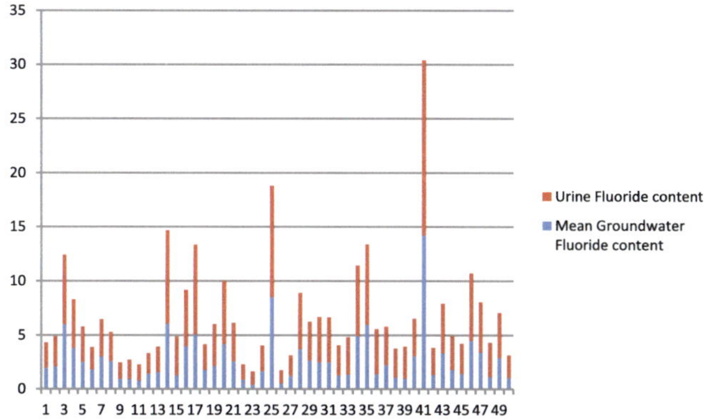

Figure 7.
Results of F− content in urine and groundwater samples of study area.*

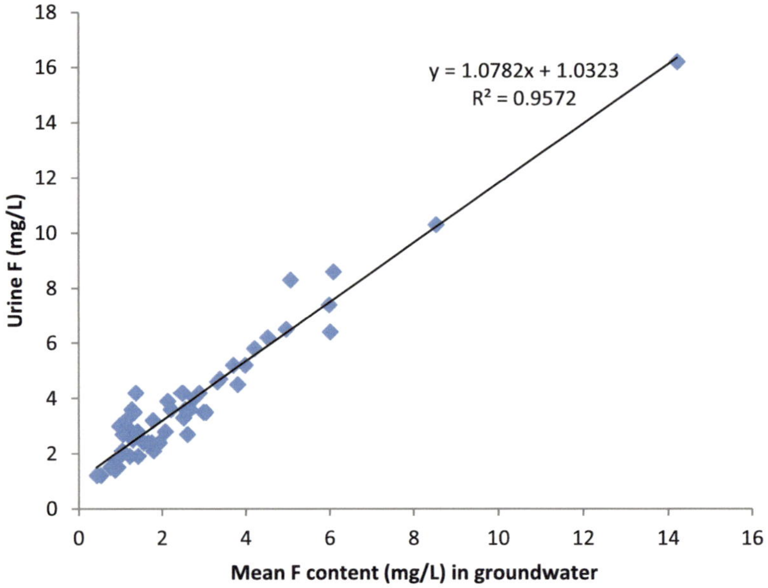

Figure 8.
Mean F– content in groundwater vs urine fluoride.

7. Conclusion

The present study reveals that the underground drinking water of the investigated area was contaminated with F. The population of the study area was therefore chronically exposed to higher levels of F from drinking water. There was a significantly positive correlation between the F content of groundwater and urine of the fluorosis-affected children in the study area. This suggested that a high level of endemic fluorosis is prevalent in the study area due to the consequence of a higher concentration of F in underground aquifers. The highest number of fluorosis-affected children (85.29%) was recorded from Datlavaripalem village. It can be concluded that the high F in groundwater leads to increased incidences of dental fluorosis among the surveyed people. Also this study indicates that urinary F is a good indicator of community exposure F. The study revealed that the F level in urine was higher than the accepted levels. It is also evident that other sources of dietary F intake other than drinking water contributed significantly to community overexposure to fluoride in the studied areas. This calls for urgent interventions to mitigate effects of excessive environmental fluoride in these areas

Acknowledgements

The author wants to thank Dr.P.V. Nageswara Rao, Assistant Professor, Department of Geology, Acharya Nagarjuna University for his constant encouragement throughout this work.

Author details

Babu Rao Gudipudi
Department of Civil Engineering, Narasaraopeta Engineering College (Autonomous), Narasaraopet, Guntur, India

*Address all correspondence to: gudipudi.br@gmail.com

References

[1] Raju NJ, Dey S, Gossel W, Wycisk P. Fluoride hazard and assessment of groundwater quality in the semi-arid Upper Panda River basin, Sonbhadra district, Uttar Pradesh, India. Hydrological Science Journal. 2012;**57**(7):1433-1452

[2] Varol M, Gokot B, Bekleyen A, Sen B. Geochemistry of Tigris River basin, Turkey; spatial and seasonal variations of major ion compositions and their controlling factors. Quaternary International. 2013;**304**:22-32

[3] Edmunds WM, Smudley PL. Fluoride in natural waters. In: Selius O, editor. Essentials of Medical Geology. London: Elsevier Academic Press; 2005. pp. 301-329

[4] Doull J, Boekelheide K, Farishia BG, Issackson RL, Klotz JB, Kumar JV. Fluoride in Drinking Water; a Scientific Review of EPA's Standards, Committee of Fluoride in Drinking Water, Board on Environmental Studies and Toxicology, Division on Earth and Life Sciences. Washington DC: National Academic Press; 2006. 530 p

[5] Janardhana Raju N. Prevalence of fluorosis in the fluoride enriched groundwater in semi-arid parts of eastern India: Geochemistry and health implications. Quaternary International. 2017;**443**(Part B):265-278. DOI: 10.1016/j.quaint.2016.05.028. ISSN: 1040-6182

[6] Grandjean P, Olsen J. Extended follow-up of cancer incidence in fluoride-exposed workers. Journal of the National Cancer Institute. 2004;**96**(10):802-803

[7] Apambire WB, Boyle DR, Michel FA. Geochemistry, genesis, and health implications of fluoriferous groundwaters in the upper Regions of Ghana. Environmental Geology. 1997;**33**:13-24. DOI: 10.1007/s002540050221

[8] Gupta SK, Deshpande RD, Agarwal M, Raval BR. Origin of high fluoride in groundwater in the North Gujarat-Cambay region, India. Journal of Hydrology. 2005;**13**(4):596-605. DOI: 10.1007/s10040-004-0389-2

[9] Liu F, Song X, Yang L, Zhang Y, Han D, Ma Y, et al. Identifying the origin and geochemical evolution of groundwater using hydrochemistry and stable isotopes in the Subei Lake basin, Ordos energy base, Northwestern China. Hydrology and Earth System Sciences. 2015;**19**:551-565. DOI: 10.5194/hessd-11-5709-2014

[10] American Public Health Association. Standard Methods for the Examination of Water and Waste Water. 21st ed. Washington, DC: APHA; 2005

[11] CGWB. Central Ground Water Board, Guntur district groundwater brochure, Andhra Pradesh Technical Report Series, 2013. pp. 1-20

[12] Sankar DB, Sai Prasad KS. Petrology of Garimanipenta (copper mineralisation area), Nellore district, Andhra Pradesh, South India—A case study. International Journal of Science and Environmental Technology. 2012;**1**(4):247-259

[13] Sesha Sai VV. Proterozoic granite magmatism along the Terrane Boundary Tectonic Zone to the east of Cuddapah basin, Andhra Pradesh—Petrotectonic implications for Precambrian Crustal Growth in Nellore Schist Belt of Eastern Dharwar Craton. Journal of Geological Society India. 2013;**81**:167-182

[14] Hem JD. Study and interpretation of the chemical characteristics of natural water. In: U S G S Professional Paper Book- 2254. Jodhpur, India: Scientific Publishers; 1991

[15] Seaber PR. Cation hydrochemical facies of groundwater in the English town Formation. New Jersey: U.S. Geol. Survey Professional Paper, 450 B; 1962. pp. 124-126

[16] Back W. Hydrochemical facies and groundwater flow pattern in northern part of Atlantic Coastal Plain. U.S. Geol. Survey Prof. Paper 498 A. 1966. p. 42

[17] Piper AM. A Graphic Procedure in the Geochemical Interpretation of Water Analysis. Washington DC: United States Geological Survey; 1953

[18] Rajiv Gandhi Drinking Water Mission. Prevention and Control of Fluorosis in India. New Delhi: Ministry of Water Resources; 1993

[19] Dahyia S, Kaur A, Jain N. Prevalence of fluorosis among school children in rural area, district Bhiwani—A case study. Indian Journal of Environmental Health. 2000;**42**:192-195

[20] World Health Organization. Guidelines for Drinking Water Quality. 2nd ed. Geneva: World Health Organization; 1997

[21] Dean HT, Elvolve E. Studies on minimal threshold of the dental sign of chronic endemic fluorosis (mottled enamel). Public Health Reports. 1935;**17**:19-29

[22] World Health Organization. Oral Health Survey-Basic Methods. 3rd ed. Geneva: World Health Organization; 1987

[23] Teresa A, Herrera M, Martin-Dominguez IR, Trejo-Vazquez R. Well water fluoride, dental fluorosis, and bone fractures in the Guadiana Valley of Mexico. Fluoride. 2001;**34**(2):139-149

[24] Augustine A, Anitha P. Health risk from fluoride exposure of a population in selected areas of Tamilnadu, South India. Food Science and Human Wellness. 2013;**2**:75-86

[25] Gibbs RJ. Mechanisms controlling world water chemistry. Science. 1970;**170**:795-840

[26] Wang YX, Shvartsev SL, Su CL. Genesis of arsenic/fluoride enriched soda water: A case study at Datong, Northern China. Applied Geochemistry. 2009;**24**:641-649

[27] Mamatha P, Rao SM. Geochemistry of fluoride rich groundwater in Kolar and Tumkur Districts of Karnataka. Environment and Earth Science. 2010;**61**:131-142

[28] Li Q, Zhou JL, Zhou YZ, Bai CY, Tao HF, Jia RL, et al. Variation of groundwater hydrochemical characteristics in the plain area of the Tarim Basin, Xinjiang Region, China. Environmental Earth Science. 2014;**72**:4249-4263

[29] Narsimha A, Sudarshan V. Contamination of fluoride in groundwater and its effect on human health: A case study in hard rock aquifers of Siddipet, Telangana State, India. Applied Water Science. 2017;7(5):2501-2512. DOI: 10.1007/s13201-016-0441-0

[30] Li P, Qian H, Wu J, Zhang Y, Zhang H. Major ion chemistry of shallow groundwater in the Dongsheng coalfield, Ordos Basin, China. Mine Water Environment. 2013;**32**:195-206

[31] Opydo-Szymaczek J, Opydo J. Fluoride content of beverages intended for infants and young children in

Poland. Food and Chemical Toxicology. 2010;**48**(10):2702-2706

[32] Opydo-Szymaczek J, Opydo J. Dietary fluoride intake from infant and toddler formulas in Poland. Food and Chemical Toxicology. 2011;**49**(8):1759-1763

[33] Raja Reddy D, Lahiri K, Ram Mohan Rao NV, Vedanayakam HS, Ebenezer LN, Suguna Ram M. Trial of magnesium compounds in the prevention of skeletal fluorosis – An Experimental Study. Fluoride. 1985;**18**(3):135-140

[34] IPCS. Fluorides. Geneva, World Health Organization, International Programme on Chemical Safety (Environmental Health Criteria 227). 2002

[35] Jaganmohan P, Narayana SVL, Sambasiva R. Prevalence of high fluoride concentration in drinking water in Nellore district, Andhra Pradesh, India: A biochemical study to develop the relation to the renal failures. World Journal of Medical Sciences. 2010;**5**:45-48

Chapter 3

Fluoride Detection and Quantification, an Overview from Traditional to Innovative Material-Based Methods

Eugenio Hernan Otal, Manuela Leticia Kim and Mutsumi Kimura

Abstract

Fluorine is the 13th most abundant element on Earth, and fluoride is part of our everyday lives, present in our drinking water, beauty products, and naturally present in food and beverages. It is a key element to increase the resistance of the dental enamel to the acidic bacteria attack and prevent dental decay. However, the ingestion of this anion for an extended period of time and in concentrations over the recommended limits can produce mild to severe health issues, called fluorosis, that can produce incorrect dental enamel formation, reduce the functionality of joints and even affect the bone structure. To avoid these terrible effects, it is necessary to control the fluoride levels in drinkable water, particularly in communities without access to safe water networks. To achieve this goal, the first step is to identify safe water sources and provision portable and reliable sensors to these communities. A major step towards safe water accessibility would be the implementation of these sensors by the proper use of new materials and technologies. Here we present an overview of the traditional quantification methodologies and the new ones for fluoride detection and quantification, and the future trends on portable devices for user-friendly on-point measurements.

Keywords: fluorides, fluorosis, sensors, metal–organic frameworks, portable devices, smartphones

1. Introduction

Fluoride is the inorganic anion of Fluorine. As all the halogens in −1 state, it generates colorless salts and can be classified as a weak base due to the $pK_a^{HF} = 3.2$. Fluorine is in the 13th position of abundance in the earth, and it is present only in the anion form as fluorite (CaF_2), the most abundant fluoride mineral.

The most relevant use for fluoride is cavity prevention, due to the presence of fluoride in water, toothpaste, and fluoride therapy in the form of sodium fluoride (NaF) or sodium monofluorophosphate (Na_2PO_4F). Water fluoridation is considered by the U.S. Centers for Disease Control and Prevention (CDC) as *"one of 10 great public health achievements of the twentieth century"* [1].

Fluoride is present in dental products, food, and drinking water. Fluoride content in dental products is between 1.0 and 1.5 mg kg^{-1}. Vegetables and fruits have a low content (0.1–0.4 mg kg^{-1}), while rice and barley can contain higher fluoride levels (2 mg kg^{-1}). Meat and fish can have higher concentrations, but it is accumulated in bones, which does not represent a risk. The dietary recommendations for adults in U.S.A. are between 3.0 and 4.0 mag day^{-1}, while in Europe are between 2.9 and 3.4 mag day^{-1}. The major known risk of fluoride deficiency is the risk of tooth cavities.

On the other hand, excess fluoride can conduce to health problems. World Health Organization (WHO) settled the recommended upper limit for fluoride in drinking water to 1.5 mg kg^{-1} [2]. Prolonged exposure to higher levels of fluorides above the recommended limit can cause dental fluorosis (1.5–3 mg kg^{-1}), which exhibits defects in enamel formation, mottling, browning, and severe teeth deterioration. Higher concentrations (4–8 mg kg^{-1}) can cause skeletal fluorosis, where the bones are hardened and less elastic, increasing the frequency of fractures. Even higher concentrations can cause crippling deformities of the spine and major joints, reducing body mobility and can also cause neurological defects and compression of the spinal cord.

The incidence of fluorosis is low in urban populations but more frequent in rural populations. The most affected areas are located in the south of South America, Southwest North America, north and east coast of Africa, India, and China [3]. In the case of east coast of Africa, fluoride concentration is related to geological formation, like volcanic activity, (East African Rift through Sudan, United Republic of Tanzania, Uganda, Ethiopia, and Kenya). Kenyan Lakes of Najura and Elmentaita presented 2800 and 1630 mg kg^{-1} fluoride, respectively, and Tanzanian Momella soda lakes presented 690 mg kg^{-1} fluoride.

As the contamination of natural waters with fluoride are mainly geogenic than anthropogenic, and thus the distribution of fluorides levels is determined by the geological formation of the riverbeds. This scenario generates an inhomogeneous distribution of fluoride levels in the water sources, even in small areas. In **Figure 1**, the distribution of fluoride levels in Arusha, Tanzania is shown. The red spots represent water sources with fluoride concentration above WHO recommendations and the blue ones below this level. The figure shows that safe and unsafe water sources can be closed and with adequate information, the local populations can choose the safer water source and avoid health risks [4].

Recently, the water fluoridation effectiveness against teeth decay was strongly questioned [5]. Countries without water fluoridation systems, like Denmark, exhibit tooth decay rates similar to US communities with fluoridation. This observation makes it necessary to rethink the need for water fluoridation to prevent cavity prevention. The amount of fluorides in toothpaste and rinses seems to be enough to protect the teeth enamel. When fluoride ions are in the mouth, they are incorporated into plaque. When the pH decreases, the fluoride ions are released from the plaque and participate in the remineralization process, which slows down the tooth decay rate. The fact that the cells involved in the remineralization process, the ameloblasts, are affected by the presence of fluorides, suggests that other cells in the body can also be affected. The relationship

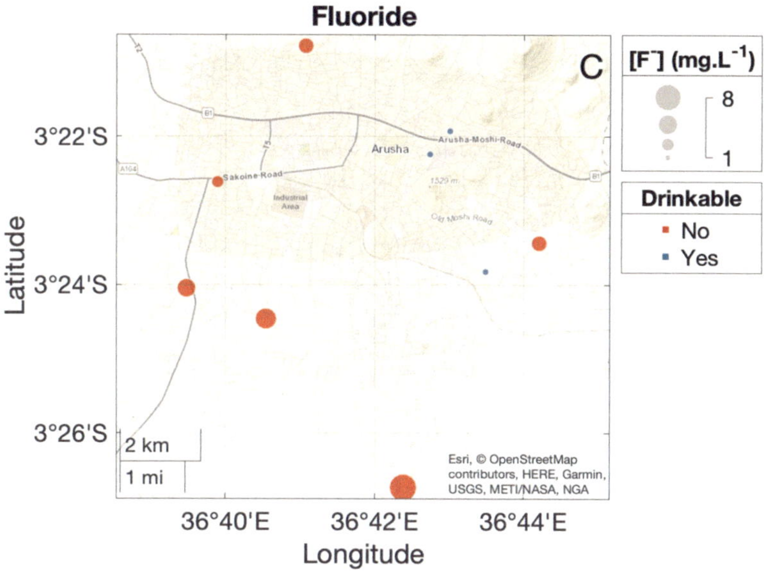

Figure 1.
Distribution of fluoride on different water sources near Arusha region. Red dots correspond to water sources with fluoride levels higher than WHO recommendations. Blue dots are water sources with fluoride levels lower than 1.5 mg kg^{-1}. Reprint with permission from ACS Sens. 2021, 6, 1, 259–266 publication date: January 8, 2021, https://doi.org/10.1021/acssensors.0c02273 copyright © 2021 American Chemical Society.

between IQ and water fluoridation was recently reported, opening the possibility of pointing to fluoride as being a developmental neurotoxin [6].

Due to severe risks on human health, it is vital to study the long-term effects of fluorides in the population, have strict control on the fluoride intake, and develop techniques to provide reliable quantification of fluoride on drinking water.

The analytical methodologies for fluoride quantification range from electrochemical approaches to colorimetric methodologies, by using naked eye detection or by means of spectrophotometric measurements. The more reliable quantification methodologies performed in laboratories, require trained operators to perform the quantification and to accurately interpret the results. However, these instruments are often out of reach for the majority of the communities in developing countries.

In order to develop accessible, reliable, and sustainable fluoride quantification methodologies, it is important to understand the chemistry involved and how new technologies like 3D printing, low-cost electronics using Arduino, and the use of smartphones as interfaces can make a major contribution to the improvement of the user experience.

In this chapter, an overview of the traditional fluoride quantification methodologies and those emerging from the use of advanced materials like Metal–Organic Frameworks will be found. Also, the implementation of smartphones as user interfaces for analytical determinations will be discussed, prioritizing the easiness, fast response, and accessibility of the methodology.

In some cases, a compromise between the accuracy or the application range will be found, but always keep in mind the convenience of the final user and the democratization of science and tech.

2. Analytical methods for fluoride determination

2.1 Electrochemical methods

The fluoride-selective electrode is a measuring electrode whose potential depends on the concentration of fluoride ions (F^-) in the solution in which it is immersed. It serves as a sensor to determine the concentration of fluoride ions. This electrode must be immersed in the solution together with a separate or built-in reference electrode so that the voltage between the electrodes can be measured with a suitable measuring instrument—some devices also convert the voltage into a concentration. A fluoride electrode can be used in a fairly wide concentration range—typically from 10^{-6} to $0.1\,\text{mol}\,\text{L}^{-1}$. Therefore, the determination method with the fluoride electrode is the most important and most frequently used for the direct determination of fluoride in drinking water. The most important part of the fluoride-selective electrode is a membrane made of a solid fluoride ion conductor, mostly a single crystal of lanthanum fluoride LaF_3, which has been doped with europium ions, Eu^{+2}. This membrane does not measure the concentration, but the activity of the fluoride ions. In order to obtain reliable measured values even for samples with fluctuating ionic strength, The sample has to be conditioned before the measurement by adding a special buffer solution (TISAB, total ionic strength adjustment buffer). This also ensures that the pH value is not too high, as hydroxide ions can interfere with the measurement. TISAB also contains reagents that react with trivalent ions, like aluminum (Al^{+3}) and iron (Fe^{+3}), forming complexes and thus preventing them from binding fluoride and thus causing a wrong fluoride determination. Then, the voltage between the fluoride-selective electrode and a reference electrode is measured. It is given according to the Nernst equation, taking into account the single negative charge of the fluoride ion

$$E = E^0 - \frac{\text{RT}}{F}.\ln \alpha_{F^-} = E^0 - \frac{\text{RT}}{F}.\ln c_{F^-} - \frac{\text{RT}}{F}.\ln \gamma_{F^-} \tag{1}$$

where

E Electrode potential measured on the fluoride electrode against the reference electrode.

E^0 Electrode potential against the same reference and with $\alpha_{F^-} = 1$.

R = 8.31447 J K^{-1} mol^{-1}.

T Absolute temperature in Kelvin: 273.15 + C^o.

F Faraday constant: 96485.34 C mol^{-1}.

a_{F^-} Activity of fluoride anions.

c_{F^-} Concentration of fluoride anions.

With the addition of TISAB, the activity coefficients keep constant and the expression simplifies to:

$$E = E^{0'} - \frac{\text{RT}}{F}\ln c_{F^-} \tag{2}$$

with

$$E^{0'} = E^0 - \frac{\text{RT}}{F}\ln \gamma_{F^-} \tag{3}$$

At constant ionic strength, pH, and 25°C, the expression is reduced to

$$E = E^{0'} - 59.2\,\text{mV}c_{F^-} \quad (4)$$

2.2 Spectroscopic methods

Spectroscopic methods are based on the high affinity of fluorides to certain metals. A colored complex can exchange its ligands with fluorides and change the color of the solution. This change can be quantified using the Lambert–Beer law using spectrophotometric measurements. For this type of determination, it is necessary to quantify the attenuation of a light source passing through a medium, in this case, the solution containing the metal complex and the fluorides. The light from a light source of Irradiance P will pass through an *infinitesimally thin* layer of the sample dx. During the light absorption process, the irradianace P will decay its power in dP, this decay will be proportional to the concentration of colored complexes c, the probability of light absorption β, and the thickness of the section dx:

$$dP = -\beta.P.c \;\; dx \quad (5)$$

The negative sign in the expression indicates that P decreases while passing through the solution.

This expression can be rearranged to:

$$-\frac{dP}{P} = -\beta.c \;\; dx \Rightarrow -\int_{P_0}^{P} \frac{dP}{P} = \beta.c \int_0^b \; dx \quad (6)$$

If we integrate this expression with limits $P = P_0$ at $x = 0$ and $P = P$ at $x = b$.

$$-\ln P - (-\ln \;\; P_0) = \beta.c.b \Rightarrow \ln \frac{P_0}{P} = \beta.c.b \quad (7)$$

Changing the logarithm base, we obtain:

$$A = \log \frac{P}{P_0} = \frac{\beta}{\ln \;\; 10}.c.b = \varepsilon.c.b \Rightarrow A = \varepsilon.c.b \quad (8)$$

This is the linear relationship between concentration and Absorbance, A, where c is the concentration of the colored analyte, b the optical path, and ε a proportional factor.

As the reaction of fluorides with a metal complex causes a change in the color intensity and this change is proportional to the fluoride concentration, the fluoride concentration can be determined using the expression $A = \varepsilon.c.b$.

The use of spectrophotometric methods to determine fluorides has a long story due to the simplicity with respect to electrochemical methods, and the most relevant will be described herein.

2.2.1 Fe-SCN system

The $Fe(SCN)^{+2}$ complex has a characteristic strong red color while the fluoride analog is colorless. The disappearance of the red color in presence of fluoride can be

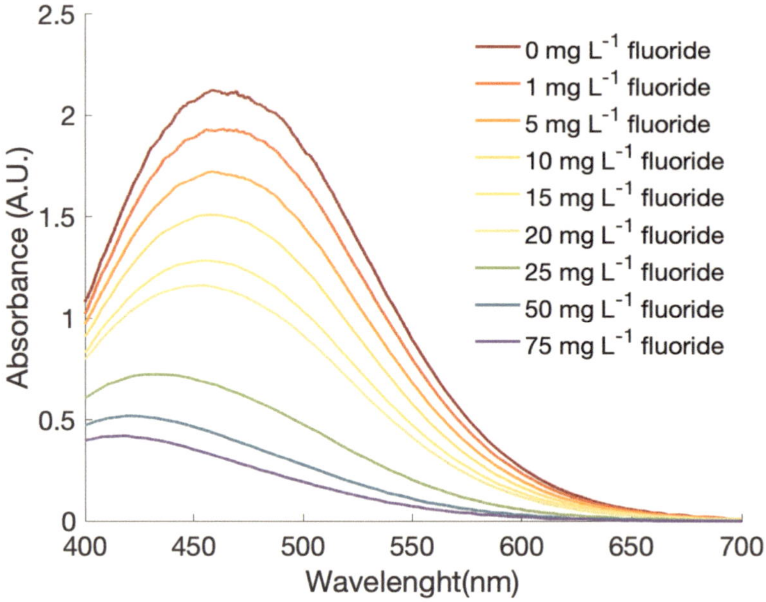

Figure 2.
Visible absorption spectra of FeSCN complex and the decrease of absorbance in presence of different concentrations of fluoride.

	FeSCN	FeF
$\log \beta_1$	2.09	5.5
$\log \beta_2$	3.30	9.7
$\log \beta_3$		12.7
$\log \beta_4$		14.9
$\log \beta_5$		15.4

Table 1.
Formation constants for Fe(III) complexes with thiocyanate and fluoride.

used for quantification purposes, and the changes in the UV–Vis spectra can be observed in **Figure 2**. The fluoride and thiocyanate complexes formation constant are shown in **Table 1**, where fluoride complexes are more favorable than thiocyanate ones and the equilibrium is displaced according to the Eq. (9).

$$Fe(SCN)_n^{3-n} + m\ F^- \leftrightarrow FeF_m^{3-m} + n\ SCN^- \tag{9}$$

Even though this method was reported in 1933 for the first time [7], it was recently implemented in a portable sensor, which achieves the WHO limits in drinking water [4]. This methodology has the advantage of being low cost, with reagents easily found in every chemistry lab. Also, the construction of the test strips using cotton as substrate, allows controlling the amount of sample used, being reproducible and user friendly. The sample enters the reaction zone by capillarity within the highly hygroscopic substrate. The quantification can be performed in two ways: (1) using photographs that the user makes from the test strips. Then the image is analyzed by splitting the signal in the Red, Green, and Blue channels (**Figure 3**). (2) using an Arduino-based device, which can be connected to a smartphone and the data is received, processed, visualized, and shared through an application [8]. Under the optimized conditions, the image analysis showed a linear range up to 15 mag L^{-1}, Relative

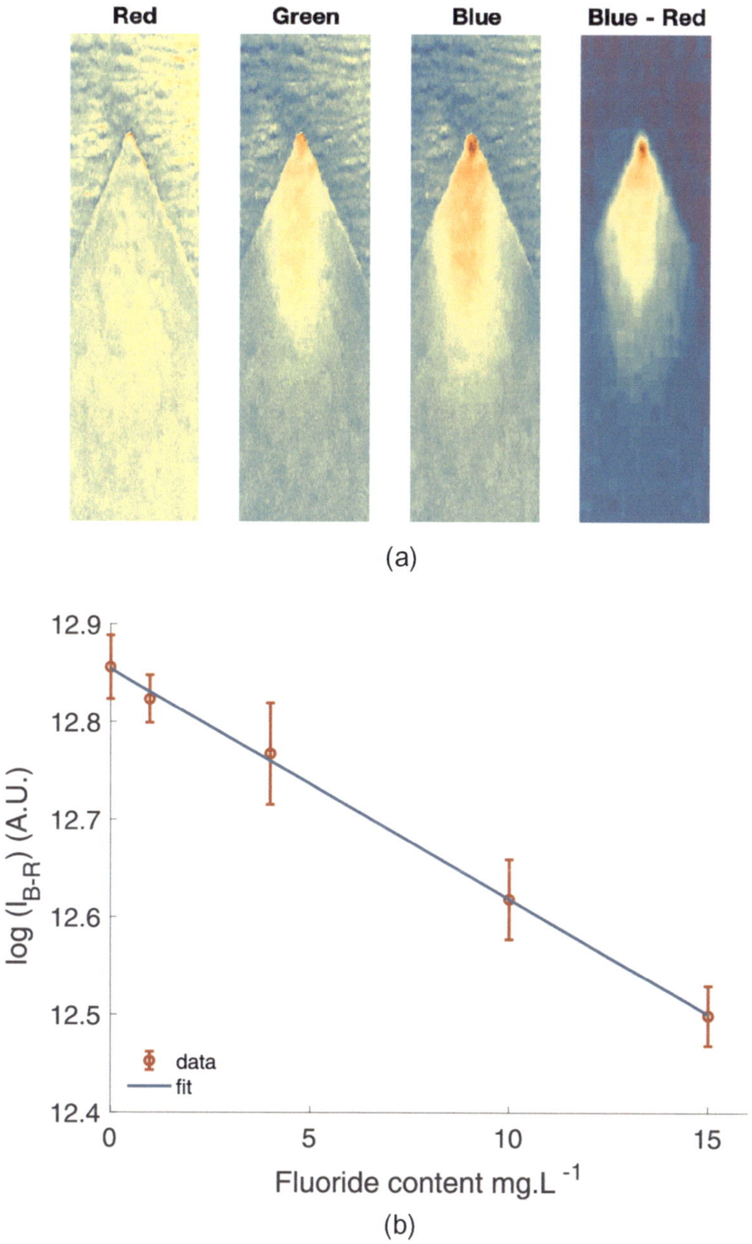

Figure 3.
(a) Images of the red, green, blue channels and the difference between blue and red components of the image. (b) Calibration curve obtained using "image analysis" quantification method using 20 μL of 0.33 mM Fe(III) in 0.4 M $HClO_4$ and 20 μL of 2.6 M SCN^- in the test strip. Sample volume approx. 260 μL. reprinted with permission from ACS Sens. 2021, 6, 1, 259–266 publication date: January 8, 2021, https://doi.org/10.1021/acssensors.0c02273 copyright Â© 2021 American Chemical Society.

Standard Deviation or RSD% of 4.3%, and a Limit of Detection or LoD of 2.8 mg L^{-1}. On the other hand, the colorimetric Arduino-based analysis showed a linear range up to 8 mg L^{-1}, RSD of 5.1%, and LoD of 0.7 mg L^{-1}. Even though the LoD values are higher than other colorimetric methodologies, the Fe-SCN methodology showed excellent recovery % even in the presence of other common anions and cations at higher concentrations than fluoride. Therefore, it is a simple, affordable yet appropriate methodology for the water quality assessment on areas where the fluoride concentration is high (e.g. United Republic of Tanzania) (**Figure 4**).

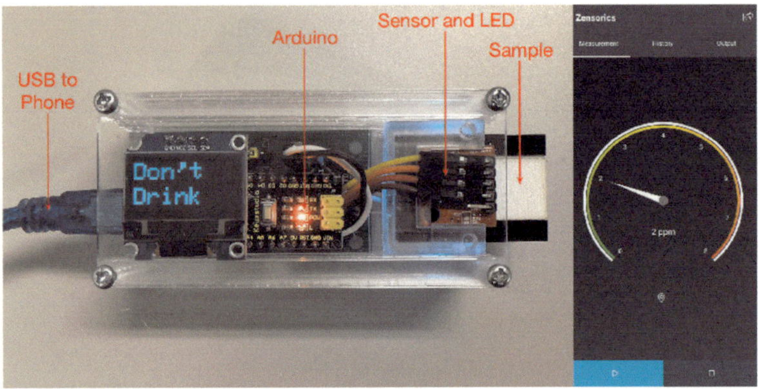

(a)

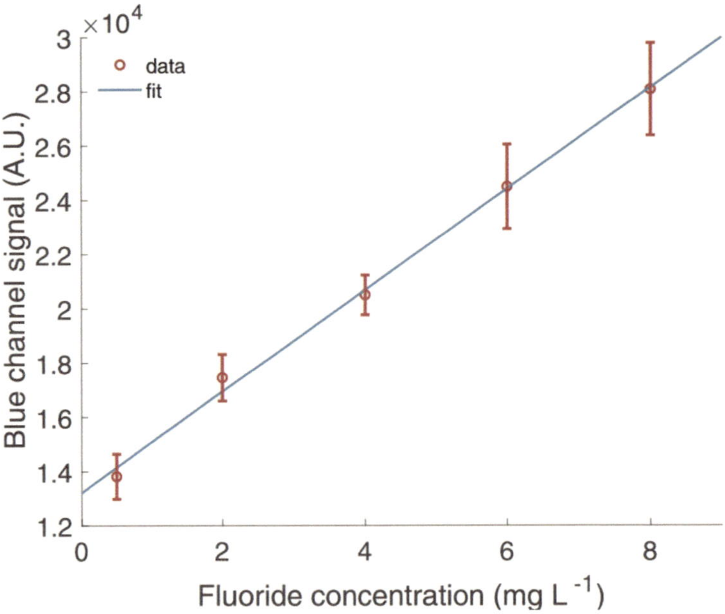

(b)

Figure 4.
(a) Photograph of the Arduino portable device for color quantification and the app developed for the visualization and data sharing (b) calibration curve obtained with the Arduino-based device. Adapted with permission from ACS Sens. 2021, 6, 1, 259–266 publication date: January 8, 2021, https://doi.org/10.1021/acssensors.0c02273 copyright © 2021 American Chemical Society.

2.2.2 Alizarin complexes

One of the most used and old methodologies for fluoride detection and quantification methodologies are those using Alizarin complexone (AC)(2-[carboxymethyl-[(3,4-dihydroxy-9,10-dioxoanthracen-2-yl)methyl]amino]acetic acid). The complex of Ce^{+3} AC in acetonitrile media gives purple complexes and the absorbance changes with fluoride can be measured at 617 nm [9]. The standardized methodology using AC accepted by the Environmental Protection Agency (USA) [10], requires the fluoride distillation from the sample prior to the measurement, increasing the probability of error due to sample manipulation and increasing the operational difficulty.

Another useful method for fluoride determination is the Zr-Alizarin S red complex. In this case, Alizarin S red or simply Alizarin (**Figure 5**) shows a yellow color in the free form and changes to red-purple complex in presence of Zr. The quantification of fluoride can be performed at 520 nm measuring the decrease of the Alizarin-Zr complex or at 425 nm, measuring the free Alizarin form freed when fluoride is present (**Figure 6**).

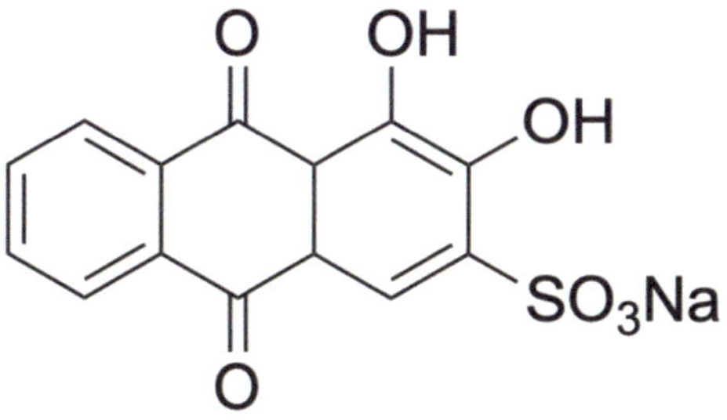

Figure 5.
Structure of 3,4-Dihydroxy-9,10-dioxo-9,10-dihydroanthracene-2-sulfonic acid, also known as alizarin S red.

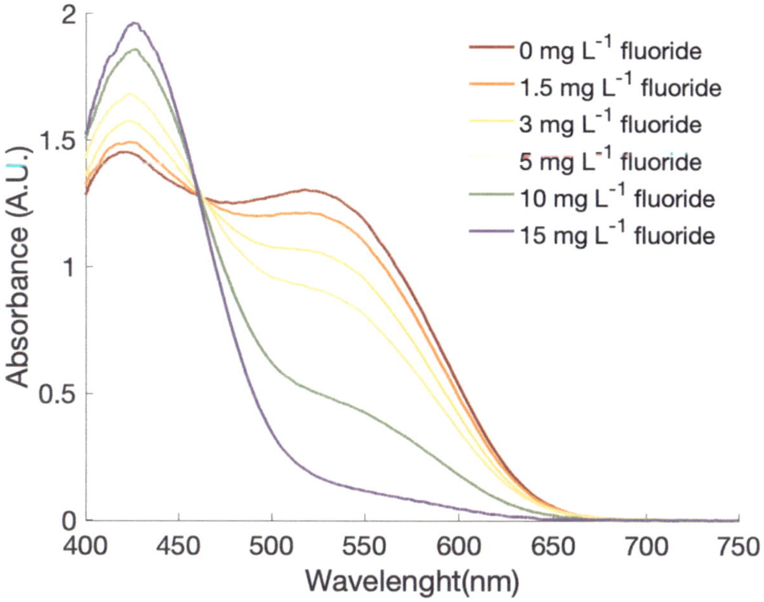

Figure 6.
Visible absorption spectra of Zr-alizarin S red complex and the decrease of absorbance in presence of different concentrations of fluoride.

2.2.3 Zr-SPADNS system

Nowadays, the accepted standardized methodology is the ion-selective methodology described above [11], but for practical reasons, many qualitative and semi-quantitative methodologies based on the colorimetric reaction of 2-(parasulfophenylazo)-1,8-dihydroxy-3,6-naphthalene-disulfonate (SPANDS, see **Figure 7**) and Zr(II) with fluorides are found [12] (see **Figure 8**). The Environmental Protection Agency uses the Zr-SPANDS methodology as their standardized methodology [13]. Commercial test strips, online methodologies [14, 15], and on-site test kits are available elsewhere showing good reproducibilities and moderately narrow linear ranges that limit their application to waters where the fluoride content is below 5 mg L^{-1}.

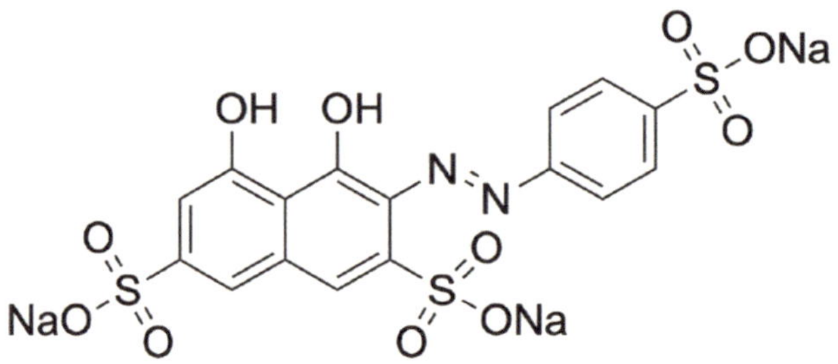

Figure 7.
2-(4-Sulfophenylazo)-1,8-dihydroxy-3,6-naphthalenedisulfonic acid trisodium salt, also known as SPADNS.

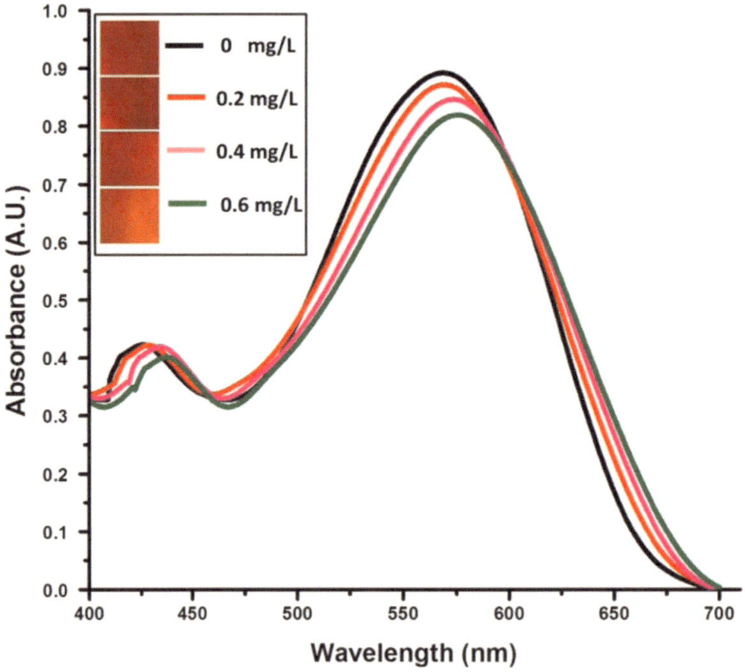

Figure 8.
Absorption spectrum of zirconium–SPADNS dye mixed with water sample containing fluoride ion at different concentrations; the inset shows the photo images of the corresponding samples. Reprinted with permission from anal. Chem. 2017, 89, 1, 767–775 publication date: December 1, 2016, https://doi.org/10.1021/acs.analchem.6b03424 copyright © 2016 American Chemical Society.

2.3 Chemosensors

As previously mentioned, the determination of fluorides in-situ is a powerful tool to provide information about the water quality in rural communities. Even though the use of ion-selective electrodes in field measurements, the simplification of spectroscopic instrumentation has far lower costs. To simplify a spectrometer is only needed to have a monochromatic light source with respect to the absorption band of the complex, a light intensity detector, and an electronic setup to read the detector output. Nowadays, this setup can be constructed using an LED as a light source, a photodiode, and a single-board microcontroller, e.g. Arduino. Also, this detector can be integrated into smartphones, providing extended capabilities.

Hussain et al. [16] reported the integration of an optical system to a smartphone (Sony Xperia E3) using Zr-SPADNS as a chemical system, ambient light sensor as a light detector, the flashlight as light source, fiber optics, and the smartphone for data collection, see **Figure 9** for optical set-up and smartphone app. Levin et al. [17] followed a similar path using zirconium xylenol orange reagent as a chemical system and using three different smartphones to test the device. Mukherjee et al. [18] used core-shell nanoparticles (near-cubic ceria@zirconia nanocages) and the same chemoresponsive dye (xylenol orange) attached to a smartphone, obtaining a linear range up to 5 mg L^{-1} and LoD = 0.1 mg L^{-1}. Otal et al. [4] use of the Fe-SNC system but instead of using the reagents in solution, the chemical were impregnated into cotton, which reduces the chemicals manipulation and provides strict control over the volume of the sample. They reported a linear range up to 8 mg L^{-1} and a LoD of 0.7 mg L^{-1}.

2.3.1 MOFs based sensors

Metal–organic Frameworks (MOFs) are a family of coordination polymers with a high surface area that can be used for water sensors among other applications. A MOF has three main points of interest:

An organic ligand, **Figure 10**, is a rigid organic molecule that is coordinated to metals and/or metallic centers. The most common coordination moiety is a carboxylate, but every moiety previously used in coordination compounds can be used here also. The ligand is the organic part and manages the isoreticular chemistry, which means that keeping unchanged the metal and the coordination moieties but changing the length of organic chain among the coordination points, the connectivity in the MOF keeps constant but the cell parameters can be expanded. An example of this is a series from UiO-66 to UiO-68, which systematically includes terephtalic acid (UiO-66), 4,4′-biphenyldicarboxylic acid (UiO-67), and p-Terphenyl-4,4″-dicarboxylic acid (UiO-68) (see **Figure 10**). Another remark about the ligand is the possibility to perform post-synthetic modification (PSM), which allows applying all the organic chemistry reactions on the synthons of these molecules (e.g. amino groups, see **Figure 10**). This toolbox is well known for many decades [19] and can be used to enhance gas storage and separation [20] and to improve photocatalytic performance [21].

A metal center, **Figure 11**. The metal center, also known as *Secondary Building Unit* (SBU), determines the topology of the MOF and also has rich chemistry due to the possibility to form *open metal sites*, which are uncoordinated metallic centers due to dangling ligands. Also the chance to include different metals in the same SBU or change the oxidation state make them an active field of research. The nature of these

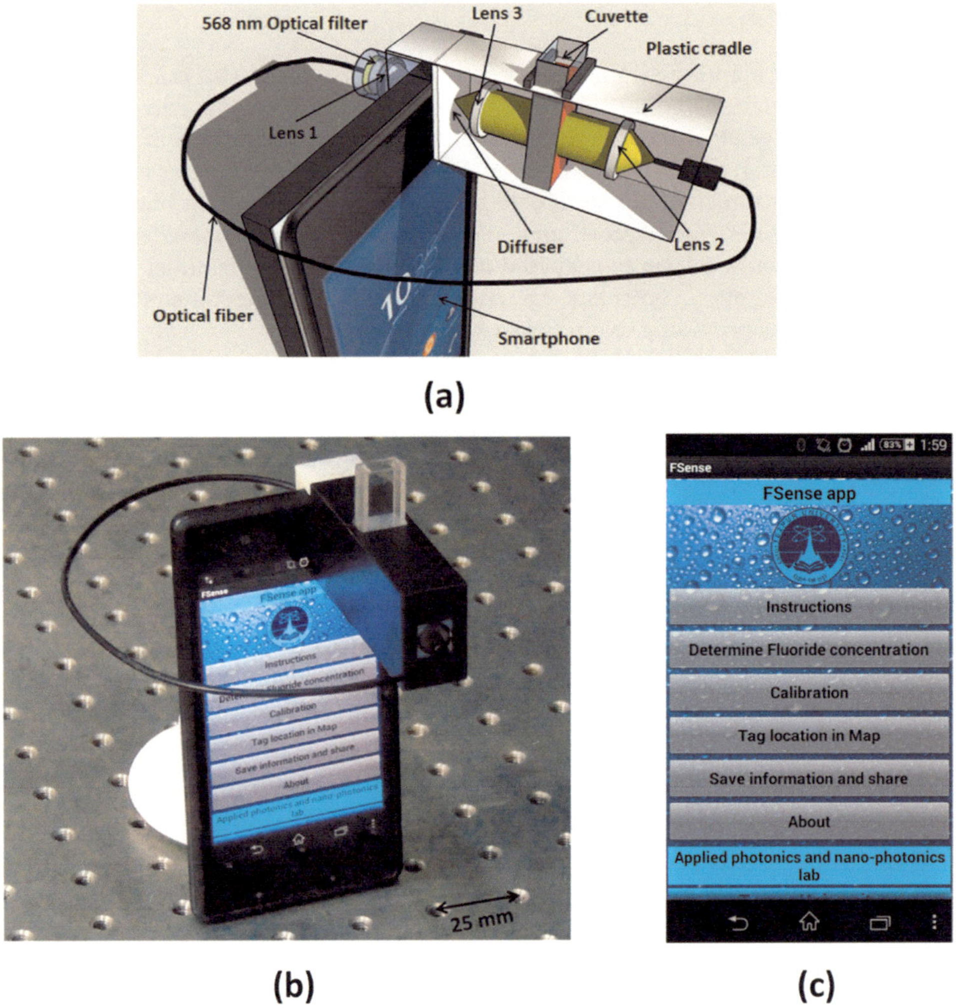

Figure 9.
(a) Schematic of the smartphone-based fluoride sensor; (b) photograph of the designed sensor; and (c) a screenshot image of the developed "FSense" application for the present sensor. Reprinted with permission from anal. Chem. 2017, 89, 1, 767–775 publication date: December 1, 2016, https://doi.org/10.1021/acs.analchem.6b03424 copyright © 2016 American Chemical Society.

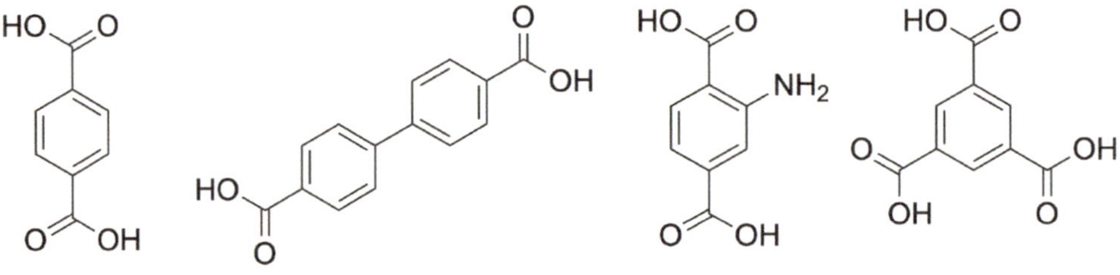

Figure 10.
Common ligands used in MOFs. From left to right: Benzene-1,4-dicarboxylic acid (terephthalic acid or BDC), Biphenyl-4,4′-dicarboxylic acid (BPDC), 2-aminobenzene-1,4-dicarboxylic acid (2-Aminoterephthalic acid), and Benzene-1,3,5-tricarboxylic acid (Trimesic acid or BTC).

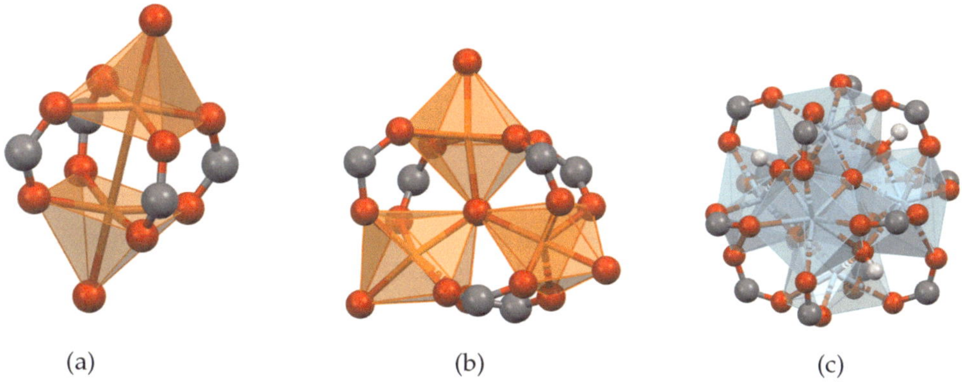

Figure 11.
Secondary building units (SBU) of selected MOFs.

centers can be from simple metal to metal–metal ligands in MOF-199 and to metal-oxo cluster like in UiO-66, see 11.

A pore, **Figure 12**. The pore is the natural consequence of the implementation of rigid ligands in a coordination polymer. The rigid ligands will keep the distance between the SBU and create a void in the center of the MOF lattice.

MOF fluoride sensors are based on the interaction of the fluorides with the SBU. This interaction is based on the affinity of the fluorides with the metal in the SBU. The metals which exhibit a strong interaction with fluorides are Al, Fe, Zr, and lanthanides. The interaction mechanism is related to the formation of a complex ion in the case of Al, Fe, and Zr, while lanthanides can form the respective fluorides, which are insoluble.

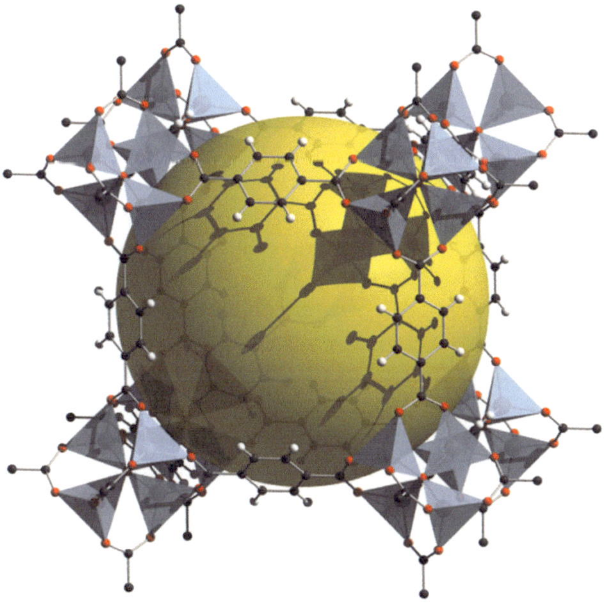

Figure 12.
Pore in MOF-5, blue tetrahedral are the Zn atoms coordinated by carboxylates and the yellow sphere represents the pore volume.

Chen and coworkers [22] used their Tb-BTC (BTC = Benzenetricarboxylates) for the fluoride detection in organic solvents like methanol and dimethylformamide. The authors found an increase of the luminescence in the presence of fluoride, with a sensing mechanism given by the confinement of the anion on the MOF's micropores, and the interaction of F^- through hydrogen bond with the solvent molecules. The intrinsic luminescence of Tb^{+3} ions is enhanced as the quenching effect of O–H bonds from the solvent is decreased.

Honjo et al. [23] used the same MOF but grown inside a liposome. And contrary to Chen, they found a decrease in the luminescence of Tb-BTC in the presence of fluoride in an aqueous buffered media (HEPES 20 mM). They found an increase in the sensitivity of these confined nanocrystals when compared with the bulk MOFs due to the enhanced dissolution of the MOF towards the formation of Tb-F non-fluorescent species. In this case, the linear range found was up to 2 mg L^{-1}.

Otal and co-workers [24] developed a portable textile-based sensor using Tb-BTC@cotton, improving the applicability of the sensor to on-site measurements of natural waters. The authors demonstrated the TbF_3 formation using synchrotron X-ray absorption fine structure measurements and proposed a 3-staged mechanism of interaction between fluorides and the luminescent MOF according to the fluoride concentration. For a given amount of solid, at low fluoride concentrations, there is an increase of luminescence with the anion concentration (ligand exchange region). These results are in agreement with the ones reported by Chen et al. [22] who also obtained an increase of the MOF luminescence with fluoride concentration. Then a "saturation" zone is observed, where the increase of fluoride concentration does not modify the luminescence intensity of the system. Finally, a "Dissolution" region appears, where the emission of the MOF decreases due to the formation of TbF_3. The cotton test-strips and the Arduino-based sensor allowed to obtain an overall low cost and easy to handle fluoride quantification system, with an extended linear range of up to 10 mg L^{-1} of fluoride, with a limit of detection of 0.8 mg L^{-1} (**Figure 13**).

Hingerholzinger and co-workers [25] used NH_2-MIL-101(Al) and fluorescein 5(6)-isothiocyanate molecules confined in the MOFs micropores. In the presence of fluorides, the MOFs dissolved releasing the dye to the media and thus, increasing the luminescence of the solution. The authors reported a linear range for fluoride of 15–1500 μg L^{-1} with a high selectivity towards the analyte, even in the presence of concomitant ions like Cl^-, Br^-, nitrates, carbonates, sulfates, and acetates. Another encapsulation of a fluorescent dye, in this case, 2′,7′-dichlorofluorescein, into the same Al-based MOF was reported by Sun et al. [26].

Zirconium-based MOFs like UiO-66 and related MOFs are highly stable in water, have high porosity, chemical, and physical stability, and a great versatility via post-synthetic modifications through the linker. These MOFs are built with $Zr_6O_4(OH)_4$ metallic centers and 1,4-Benzenedicarboxylates (BDC) as organic ligands, but they can be changed by NH_2-BDC or other functional groups.

Zhu and co-workers [27] used NH_2-UiO-66 for fluoride sensing and quantification in waters. The mechanism proposed by the authors relies on the hydrogen bond formation between the fluoride and the amino groups of the linkers. The withdrawal of electronic density away from the metallic center produces an increase in the luminescence, with a linear range up to 50 mg L^{-1} and a LoD of 0.229 mg L^{-1} of fluoride, even in the presence of common concomitants.

Also, UiO-66 MOFs were used as host frameworks for fluorescent guests within their structure. Inorganic guests like Tb^{+3} were tested for fluoride detection by

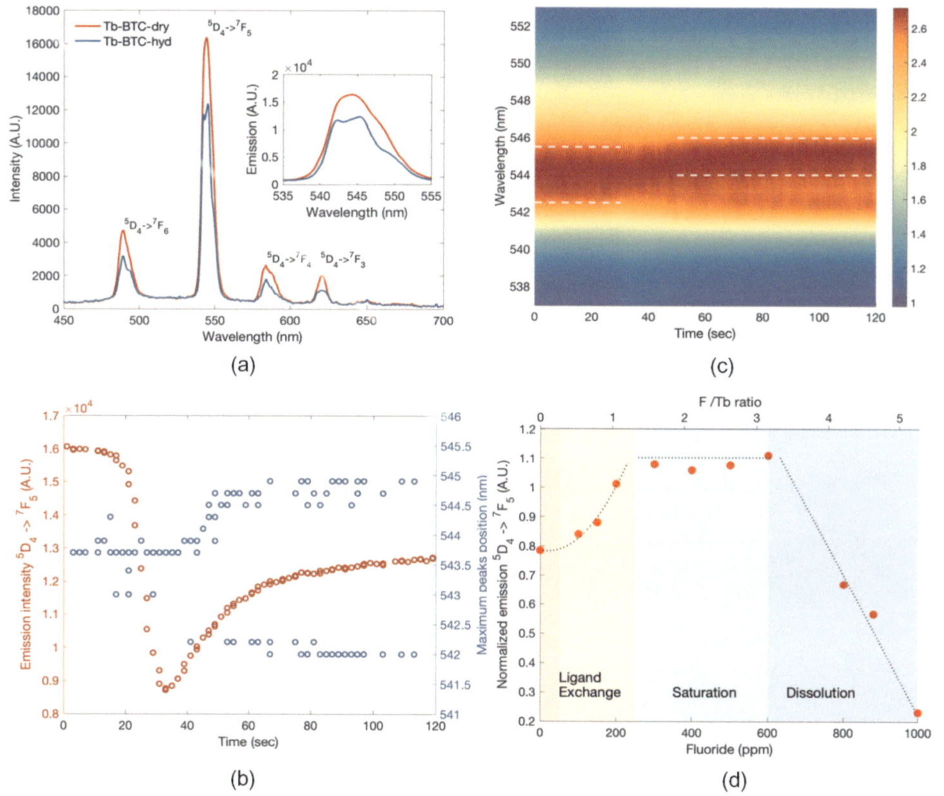

Figure 13.
(a)Emission spectra of TbBTC modified cotton before and after water exposure. (b) Intensity and maximum signal position sample are in contact with water. (c) Normalized intensity in function of time when the sample is in contact with water. (d) Intensity in function of F to Tb ratios. Regions of the proposed mechanism. Adapted with permission from [24].

incorporating open metal sites on the MOF through the partial substitution of BDC linker (**Figure 12**) with isopthalates [28]. The uncoordinated carboxy groups incorporate Tb(III) via post-synthesis which conferred a strong luminescence to the final solid. The MOFs were tested for fluoride detection, which enhanced the luminescence of the MOF and other anions like Cl^-, Br, NO_3^-, CO_3^{-2}, HCO_3^-, SiO_3^{-3}, SO_4^{-2}, and PO_4^{-3} produced a slight decrease of luminescence, and I^-, S^{-2}, and NO_2^- gave a total quenching of the MOF. The linear range was up to 40 mg L^{-1} and a LoD of 0.35 mg L^{-1}.

On the other hand, organic fluorescent guests (like fluorescein sodium) were used on UiO-66 MOFs structure [29]. The dissolution of the MOF and thus, the release of the fluorescent probe was proposed as a sensing mechanism, with a linear range up to 7.6 mg L^{-1} and a LoD of 0.08 mg L^{-1}.

Recently our group developed a simple post-functionalization procedure for Al-BDC MOFs through a thermal treatment [30] opening the possibilities towards new MOFs for fluoride sensing (**Figure 14**).

Several xanthene dyes were used as modifiers (i.e. Fluorescein, Rhodamine B, Eosin Y, Erythrosine B, and Rose Bengal) and the dissolution of the MOF in presence of fluoride and the release of the dye to the solution was measured.

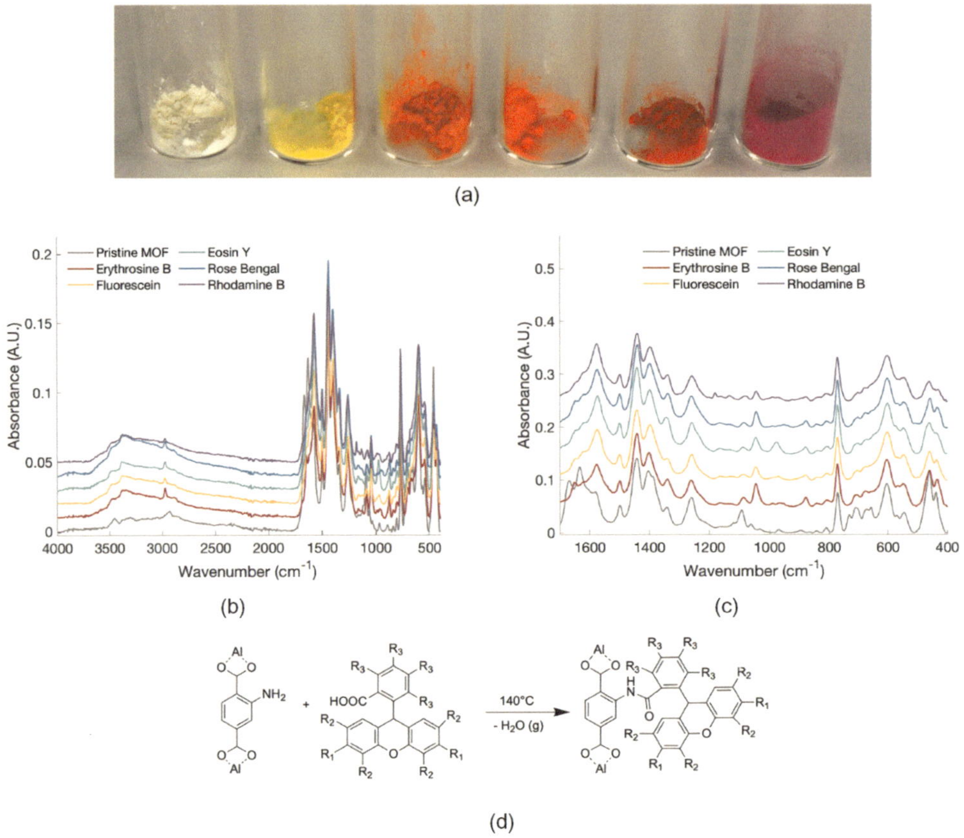

Figure 14.
*(**a**) Photograph of the modified MOF, from left to right: Al-BDC-NH$_2$, Fluorescein, Eosin Y, Erythrosine B, Rose Bengal and rhodamine B. FTIR spectra from the modified and unmodified MOFs, (**b**) complete spectra and (**c**) carboxylate region. Adapted with permission from [30]. Adapted from [30] (**d**) proposed synthetic pathway for the formation of the amide. For fluorescein: R_1 = -OH, R_2 =R_3= –H, rhodamine B: R_1 = -N(et)$_2$, R_2 = R_3 = –H, rose bengal: R_1 = -OH, R_2 = –I, R_3 = –Cl, eosin Y: R_1 = –OH, R_2 = -Br, R_3 = –H, and erythrosine B: R_1 = –OH, R_2 = –I, R_3 = –H. Adapted with permission from [30].*

Excellent recovery% was obtained even in the presence of common concomitants in waters, with a sensing mechanism governed by ligand exchange and dye release to the aqueous media (Eqs. (10)–(14)).

$$\sim \mathrm{Al} - \mathrm{OH} + \mathrm{F}^- \rightarrow \sim \mathrm{Al} - \mathrm{F} + \mathrm{OH}^- \quad (10)$$

$$\sim \mathrm{Al} - \mathrm{L}_A - \mathrm{Al} + \mathrm{F}^- + \mathrm{H_2O} \rightarrow \sim \mathrm{Al} - \mathrm{F} + \sim \mathrm{Al} - \mathrm{L}_A\mathrm{H} + \mathrm{OH}^- \quad (11)$$

$$\sim \mathrm{Al} - \mathrm{L}_A\mathrm{H} + \mathrm{F}^- + \mathrm{H_2O} \rightarrow \sim \mathrm{Al} - \mathrm{F} + \mathrm{L}_A\mathrm{H}_2 + \mathrm{OH}^- \quad (12)$$

$$\sim \mathrm{Al} - \mathrm{L}_B - \mathrm{Al} \sim + \mathrm{F}^- + \mathrm{H_2O} \rightarrow \sim \mathrm{Al} - \mathrm{F} + \sim \mathrm{Al} - \mathrm{L}_B\mathrm{H}^+\mathrm{OH}^- \quad (13)$$

$$\sim \mathrm{Al} - \mathrm{L}_B\mathrm{H} + \mathrm{F}^- + \mathrm{H_2O} \rightarrow \sim \mathrm{Al} - \mathrm{F} + \mathrm{L}_B\mathrm{H}_2 + \mathrm{OH}^- \quad (14)$$

3. Future perspectives

The chemosensors and MOF-based sensors for fluoride showcased in this chapter showed varied opportunities for naked-eye or instrumental-based colorimetric and fluorometric detection.

However, most of the sensors showed limited linear ranges and relatively high LoD values. According to current international regulations, these sensors could be of interest for their applications in rural areas and for low-cost devices. But if the levels of suggested upper limits of fluorides drop to lower values than 1.5 mg L^{-1}, then the sensitivity of these sensors should be improved. The application of Metal–Organic Frameworks for fluoride sensing is a growing area of research and it is expected the development of new materials with lower cost and better performance in the next years.

Current commercial test kits are semi-quantitative or qualitative methodologies and still rely on the human eye for reading and interpretation. They also need the reagents handling, with short shelf lives that might lead to errors and biased results that could affect human health at different levels. Therefore, the implementation of mobile phones as user-friendly devices for the quantification, monitoring, and data sharing platforms might allow the extended reach of new materials and technologies to the final users, giving accurate unbiased results, at low costs and easy sample handling.

Appendix A: common quality parameters for the analytical methodologies

Linear range Is the analyte concentration range where the response of the analytical methodology is linear and proportional to the amount of the target analyte.

Limit of Detection Is the amount of analyte giving a *significative different* signal from the blank sample. With significative differences, the statistical meaning is considered.

$$Y_{\mathrm{LoD}} = Y_{\mathrm{blank}} + 3\mathrm{S} \tag{15}$$

where Y_{LoD} is the signal of the minimum amount of analyte, Y_{blank} is the signal given by the blank sample and then, the LoD from the calibration curve can be calculated simply by

$$\mathrm{LoD} = 3\mathrm{S}/\mathrm{m} \tag{16}$$

being m, the sensitivity of the analytical methodology or the slope of the linear regression.

Accuracy Indicates how close to the *real value* is the results obtained with the analytical methodology. The evaluation of the method accuracy can be performed by using a certificate material, spiking a real sample, comparing the developed methodology with other analytical techniques, etc.

Relative Standard Deviation Gives a measurement of the precision of the methodology, especially useful when it has to be compared with other analytical methodologies available for the same analyte. It can be calculated by

$$\mathrm{RSD} = 100 * S/\overline{x} \tag{17}$$

where $\overline{x}$ is the mean value of N replicates.

Author details

Eugenio Hernan Otal[1†], Manuela Leticia Kim[1†] and Mutsumi Kimura[1,2,3*]

1 Faculty of Textile Science and Technology, Department of Chemistry and Materials, Shinshu University, Ueda, Nagano, Japan

2 COI Aqua-Innovation Center, Shinshu University, Ueda, Nagano, Japan

3 Research Initiative for Supra-Materials, Shinshu University, Ueda, Nagano, Japan

*Address all correspondence to: mkimura@shinshu-u.ac.jp

† These authors contributed equally.

References

[1] Gooch BF. Community water fluoridation—One of the 10 greatest public health achievements of the 20th century. Blogs|CDC. 2015;**130**(4): 296-298. Available from: https://blogs.cdc.gov/pcd/2015/04/23/community-water-fluoridation-one-of-the-10-greatest-public-health-achievements-of-the-20th-century/

[2] Organization WH. Guidelines for Drinking-water Quality. Geneva Switzerland: World Health Organization; 1993

[3] Yeung CA. A systematic review of the efficacy and safety of fluoridation. Evidence-Based Dentistry. 2008;**9**(2): 39-43. Available from: https://www.nature.com/articles/6400578

[4] Otal EH, Kim ML, Dietrich S, Takada R, Nakaya S, Kimura M. Open-source portable device for the determination of fluoride in drinking water. ACS Sensors. American Chemical Society. 2021;**6**(1):259-266. DOI: 10.1021/acssensors.0c02273

[5] Gravitz L. The fluoride wars rage on. Nature. 2021. DOI: 10.1038/d41586-021-02924-6. Available from: https://www.nature.com/articles/d41586-021-02924-6

[6] Grandjean P, Hu H, Till C, Green R, Bashash M, Flora D, et al. A benchmark dose analysis for maternal pregnancy urine-fluoride and IQ in children. medRxiv. 2020;**10**(31):20221374. Available from: https://www.ncbi.nlm.nih.gov/pmc/articles/PMC7654913/

[7] Foster MD. Colorimetric determination of fluoride in water using ferric chloride. Industrial & Engineering Chemistry Analytical Edition. American Chemical Society. 1933;**5**(4):234-236. DOI: 10.1021/ac50084a005

[8] Dietrich S. Zensorics App. Available from: https://hello.fridie.de/zensorics-app/

[9] Yamamura SS, Wade MA, Sikes JH. Direct Spectrophotometric Fluoride Determination. Analytical Chemistry. American Chemical Society. 1962; **34**(10):1308-1312. DOI: 10.1021/ac60190a033

[10] EPA. Methods for Chemical Analysis of Water and Wastes. Cincinnati, OH: Environmental Protection Agency; 1979. Available from: https://www.osti.gov/biblio/6259902-methods-chemical-analysis-water-wastes

[11] ASTM International. ASTM D1179-16—standard test methods for fluoride ion in water. Engineering; **360**:2-4. Available from: https://standards.globalspec.com/std/3861620/astm-d1179-16

[12] Marier JR, Rose D. The fluoride content of some foods and beverages–a brief survey using a modified Zr-SPADNS method. Journal of Food Science. 1966;**31**(6):941-946. DOI: 10.1111/j.1365-2621.1966.tb03273.x. Available from: https://onlinelibrary.wiley.com/doi/abs/10.1111/j.1365-2621.1966.tb03273.x

[13] US EPA O. Method 13A—Total Fluoride—SPADNS Zirconium Lake [Other Policies and Guidance]. 2016. Available from: https://www.epa.gov/emc/method-13a-total-fluoride-spadns-zirconium-lake

[14] Arancibia JA, Rullo A, Olivieri AC, Nezio SD, Pistonesi M, Lista A, et al. Fast spectrophotometric determination of fluoride in ground waters by flow injection using partial least-squares

calibration. Analytica Chimica Acta. 2004;**512**(1):157-163. Available from: https://www.sciencedirect.com/science/article/pii/S0003267004002168

[15] Marques TL, Coelho NMM. Proposed flow system for spectrophotometric determination of fluoride in natural waters. Talanta. 2013; **105**:69-74. Available from: https://www.sciencedirect.com/science/article/pii/S0039914012010247

[16] Hussain I, Ahamad KU, Nath P. Low-cost, robust, and field portable smartphone platform photometric sensor for fluoride level detection in drinking water. Analytical Chemistry. American Chemical Society. 2017;**89**(1): 767-775. DOI: 10.1021/acs.analchem.6b03424

[17] Levin S, Krishnan S, Rajkumar S, Halery N, Balkunde P. Monitoring of fluoride in water samples using a smartphone. The Science of the Total Environment. 2016;**551-552**:101-107

[18] Mukherjee S, Shah M, Chaudhari K, Jana A, Sudhakar C, Srikrishnarka P, et al. Smartphone-based fluoride-specific sensor for rapid and affordable colorimetric detection and precise quantification at sub-ppm levels for field applications. ACS Omega. American Chemical Society. 2020;**5**(39): 25253-25263. DOI: 10.1021/acsomega.0c03465

[19] Kalaj M, Cohen SM. Postsynthetic modification: An enabling technology for the advancement of metal–organic frameworks. ACS Central Science. American Chemical Society. 2020;**6**(7): 1046-1057. DOI: 10.1021/acscentsci.0c00690

[20] Li B, Wen HM, Zhou W, Chen B. Porous metal–organic frameworks for gas storage and separation: What, how, and why? The Journal of Physical Chemistry Letters. American Chemical Society. 2014;**5**(20):3468-3479. DOI: 10.1021/jz501586e

[21] Otal EH, Kim ML, Calvo ME, Karvonen L, Fabregas IO, Sierra CA, et al. A panchromatic modification of the light absorption spectra of metal–organic frameworks. Chemical Communications. Royal Society of Chemistry. 2016; **52**(40):6665-6668. Available from: https://pubs.rsc.org/en/content/articlelanding/2016/cc/c6cc02319c

[22] Chen B, Wang L, Zapata F, Qian G, Lobkovsky EB. A luminescent microporous metalâ^'organic framework for the recognition and sensing of anions. Journal of the American Chemical Society. 2008;**130**(21): 6718-6719. DOI: 10.1021/ja802035e

[23] Honjo M, Koshiyama T, Fukunaga Y, Tsuji Y, Tanaka M, Ohba M. Sensing of fluoride ions in aqueous media using a luminescent coordination polymer and liposome composite. Dalton Transactions. The Royal Society of Chemistry. 2017;**46**(22):7141-7144. Available from: https://pubs.rsc.org/en/content/articlelanding/2017/dt/c7dt01071k

[24] Otal EH, Tanaka H, Kim ML, Hinestroza JP, Kimura M. The long and bright path of a lanthanide MOF: From basics towards the application. Chemistry—A European Journal. 2021; **27**(26):7376-7382. DOI: 10.1002/chem.202005222 Available from: https://onlinelibrary.wiley.com/doi/abs/10.1002/chem.202005222

[25] Hinterholzinger FM, Rühle B, Wuttke S, Karaghiosoff K, Bein T. Highly sensitive and selective fluoride detection in water through fluorophore release from a metal-organic framework.

Scientific Reports. 2013;**3**(1):2562. Available from: https://www.nature.com/articles/srep02562

[26] Sun Y, Xu X, Zhao Y, Tan H, Li Y, Du J. Luminescent metal organic frameworks–based chemiluminescence resonance energy transfer platform for turn–on detection of fluoride ion. Talanta. 2020;**209**:120582. Available from: https://www.sciencedirect.com/science/article/pii/S0039914019312159

[27] Zhu H, Huang J, Zhou Q, Lv Z, Li C, Hu G. Enhanced luminescence of NH2-UiO-66 for selectively sensing fluoride anion in water medium. Journal of Luminescence. 2019;**208**:67-74. Available from: https://www.sciencedirect.com/science/article/pii/S0022231318315436

[28] Zheng HY, Lian X, Qin SJ, Yan B. Novel "turn-on" fluorescent probe for highly selectively sensing fluoride in aqueous solution based on Tb3+-functionalized metal–organic frameworks. ACS Omega. American Chemical Society. 2018;**3**(10):12513-12519. DOI: 10.1021/acsomega.8b02134

[29] Zhao X, Wang Y, Hao X, Liu W. Fluorescent molecule incorporated metal-organic framework for fluoride sensing in aqueous solution. Applied Surface Science. 2017;**402**:129-135. Available from: https://www.sciencedirect.com/science/article/pii/S0169433217300764

[30] Otal EH, Kim ML, Hattori Y, Kitazawa Y, Hinestroza JP, Kimura M. Versatile covalent postsynthetic modification of metal organic frameworks via thermal condensation for fluoride sensing in waters. Bioengineering. 2021;**8**(12):196. Available from: https://www.mdpi.com/2306-5354/8/12/196

Chapter 4

Water Defluoridation Methods Applied in Rural Areas over the World

Enos Wamalwa Wambu, Franco Frau, Revocatus Machunda, Lilliane Pasape, Stephen S. Barasa and Giorgio Ghiglieri

Abstract

Overexposure to fluoride (F) through drinking water is the most widespread water problem in the world, but it has now exacerbated due to rapid population growth rates, adverse climatic changes, and increasing levels of water scarcity. Thus, despite the large amounts of data, which has accrued on mitigation methods of high F is still the primary impediment to drinking water programs among many developing nations. The current review chapter on F mitigation techniques applied world-over is aimed at providing a succinct overview of water defluoridation techniques and strategies being used to combat the impact of human F overexposure. It represents a starting point to understand the prospects of reducing the global F impact. It is anticipated that this work will lay a strong foundation for this and also inform strategies for safeguarding public health and the environment from F pollution.

Keywords: defluoridation technologies, drinking water, fluoride, fluorosis, literature review

1. Introduction

The beneficial and detrimental effects of fluoride (F) were established in the early 1940s. Low levels of drinking water F (< 0.1 mg/L) were linked to the occurrence of dental caries, whereas elevated levels of F in water were associated with incidences of dental fluorosis among the communities [1]. Then some countries began artificial fluoridation of drinking water to control teeth decay [2]. Soon the widespread use of F in drinking water and oral products to control teeth decay resulted in a drastic decline in incidences of dental caries with a concomitant rise in dental fluorosis among the communities [3]. The severity of fluorosis increases with increasing F concentrations greater than 1.5 mg/L in drinking water [4] and data over the last few decades indicate trends toward more fluorosis around the world [5].

The new surge in the prevalence of fluorosis around the world has been attributed to, among other the rise in water fluoridation programs [5]; indiscriminate use of fluoridated products [6]; inadequate F legislation in the affected countries [7]; lack of technology and capacity for sustainable F surveillance [2] and widespread water security problems [8]. The WHO's "guidelines" of 1.5 mg/L as the allowable standards of drinking water F have also come under scrutiny in regard to its success in controlling the adverse F effects among the communities [9]. Severe fluorosis has been recorded among communities using household waters with F levels well within these guidelines [10]. So, there have been efforts to control the adverse F effects in the communities. The strategies that are employed include instituting community health risk assessments and management programs [11]; high water F surveillance [12]; prospecting for safer water [13]; development of water defluoridation strategies; F awareness creation and behavior change campaigns [14]; developing water policies and legislation for F mitigation [9].

It is apparent, however, that there is an urgent need to reconsider the current approaches. The desired approaches for effective control of F impact among the communities should not only be effective but also be affordable, holistic, and comprehensive. In the current work, a review of the previous strategies and methodologies that have been applied in water defluoridation is presented [15]. The chapter aims to provide an update on F mitigation technologies being deployed worldwide and it is expected that this will enhance scientific understanding of the available technologies and the prospect of reducing the global F impact.

Defluoridation of existing waters is the main option where alternate safe water in high F areas is not available. However, the available water defluoridation approaches differ in scale, efficacy, sustainability, affordability, and acceptability. Therefore, the security of supply is heterogeneous. In general, the available options for water defluoridation can be classified as chemical, membrane-based, physical, or adsorption-based.

2. Chemical methods for water defluoridation

Chemical methods of water defluoridation involve the addition of reactive species, which can react with and facilitate water F removal through phase separation steps [16]. Here, five methods of which two are based on precipitation, two on coagulation, and one on electrocoagulation processes will be explained in further detail.

2.1 Precipitation processes

They basically include lime-softening and contact precipitation. The former is a precipitation chemical method of water defluoridation used to remove calcium and magnesium ions from hard water and it is also used to reduce F levels in potable waters [17]. In the traditional "lime-softening" technique, lime ($Ca(OH)_2$) reacts with soluble $Ca(HCO_3)_2$ to precipitate insoluble $CaCO_3$ [18] according to Eq. (1) as follows:

$$Ca(HCO_3)_2(aq) + Ca(OH)_2(aq) \rightarrow 2CaCO_3(s) + 2H_2O(l) \quad (1)$$

In presence of water F, however, part of $Ca(OH)_2$ precipitates and removes insoluble fluorite, CaF_2, according to Eq. (2) as:

$$Ca(OH)_2(aq) + 2F^-(aq) \rightarrow CaF_2(s) + OH^-(aq) \quad (2)$$

The removal of F by lime-softening is enhanced in presence of dissolved magnesium salts, which precipitate as $Mg(OH)_2$ according to Eq. (3) below [19].

$$Mg^{2+}(aq)+2OH^{-}(aq)\rightarrow Mg(OH)_2(s) \quad (3)$$

This also helps to eliminate excess alkalinity in treated water [20]. Defluoridation by lime is achieved by surface precipitation of fluoride onto the $Mg(OH)_2$ formed but the process is temperature dependent and the F adsorption onto $Mg(OH)_2$ increases at high temperatures.

Lime softening is the most common water defluoridation method in developing countries [21]. Based on this technique, the Indian Institute of Science in Bangalore, developed a simple defluoridation technique, which uses magnesium oxide, lime, and sodium bisulphate [22]. Due to the presence of MgO, the pH of treated water has to be adjusted to desirable levels (6.5 to 8.5) by adding 0.15 to 0.2 g per liter of sodium bisulphate. Even so, water defluoridation based on lime softening is inefficient and the technique requires large amounts of reagents leading to high volumes of F-laden sludge.

2.2 Contact precipitation

Contact precipitation employs the simultaneous addition of soluble calcium and phosphate compounds to brackish water. These react with F ions to precipitate CaF_2 and fluorapatite ($Ca_5(PO_4)_3F_2$) [23] catalyzed by a saturated bone char medium. The process has been applied in Tanzania on raw water with 13 mg/L F resulting in F removal efficiency of 97.9% [24]. Contact precipitation has been floated as being more efficient and reliable than lime-softening, but it also generates large volumes of F-enriched sludge [25] and it can impart bad taste and smell to the treated water compromising its palatability.

2.3 Coagulation techniques

Coagulation is a procedure in which soluble metal cations [26] or commercial polyelectrolytes [27] with a large charge-to-volume ratio are added to water to attract and react with organics and other insoluble aggregates to flocculate and sediment and phase them out of the water. The flocculates formed to provide high sorbent surfaces for F ions. Alum ($Al_2(SO_4)_3.18H_2O$) is the usual flocculant in cases where F removal is also desired [26]. The salt reacts with OH^- ions to form $Al(OH)_3$ flocs according to Eq. (4) as:

$$Al_2(SO_4)_3\cdot 14H_2O+3Ca(HCO_3)_2\rightarrow 2Al(OH)_3+3CaSO_4+14H_2O+6CO_2 \quad (4)$$

The flocs sorb F from the water. A little lime is added controllably to replenish the OH^- ions. The amount of alum is controlled to prevent the initiation of complexation of Al^{3+} with F. Also, coagulation is not efficient and complete F removal is not achieved. Large amounts of coagulants are used leading to large volumes of sludge [25] and the residual coagulants in the water must be monitored to meet the drinking water standards.

Nonetheless, several authors have reported the application of this method with varying degrees of success [28, 29]. On the other hand, some authors have reported investigations aimed at improving upon the technique. Atia et al. [30], for example, compared the coagulants and demonstrated that the use of $Al_2(SO_4)_3.18H_2O$ as flocculants was superior to $Fe_2(SO_4)_3.H_2O$. However, F removal by coagulation using poly-aluminum chloride (PAC) has also been reported [31] and compared with other polyelectrolyte coagulants [32]. Most recently some workers have utilized inorganic polymeric coagulant and indicated that 80% defluoridation of 6 mg/L F polluted water [33]. An alternate precipitation technique based on induced crystallization under extreme pH levels and carbonate/bicarbonate equilibriums has recently been applied with high F removal efficiencies [30]. Also, a facile approach to calcium co-precipitation has been reported [34].

2.4 Nalgonda technique

The Nalgonda technique is a modified coagulation protocol, which was developed by National Environmental Engineering Research Institute, Nagpur, India [35]. It takes advantage of the synergy between precipitation using lime and alum flocculation to facilitate F adsorption on $Al(OH)_3$ flocs formed in the solution [36]. The F adsorbed is removed with the flocs by sedimentation. The Nalgonda technique has now been introduced in many countries but it requires high doses of flocculants leading to large sludge transfers and it may release excess Al^{3+} into the treated water [37].

2.5 Electrocoagulation

Electrocoagulation (EC) uses an electrolytic cell (**Figure 1**) [38] to supplies coagulant Al^{3+} ions in a controlled manner [26].

As the electric current passes through the cell, the Al anode gets oxidized to Al^{3+} ions, which are transformed into polymeric species and reacted with hydroxyl ions in solution to form the $Al(OH)_3$ flocs, facilitate F removal from water as in the other coagulation techniques [39].

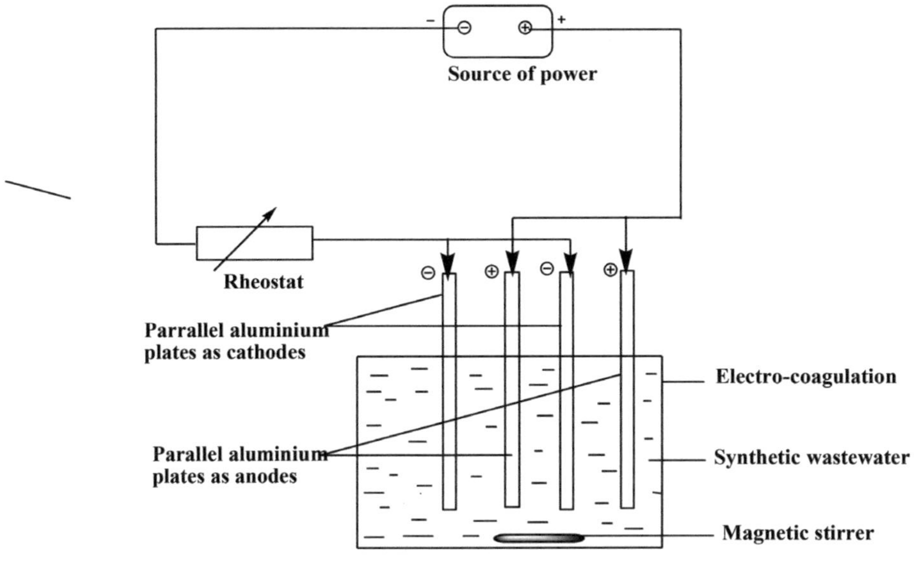

Figure 1.
Schematic representation of an electrocoagulation cell (adapted from Ref. [38]).

Even though the EC technique has been associated with demerits including high running costs, need for reliable electric power, and for specialized personnel to operate [25], the technique is efficient, reliable, and produces high-quality water. The sludge generation is low, and the method is used in different water conditions. In Algeria [40], for example, EC was applied as an efficient affordable water defluoridation technique, and, elsewhere, it has been used to reduce borehole F level from 3.5–4.8 to 0.8–1.0 mg/L [41]. Also, Khatibikamala et al. [42] reported that F concentration was reduced, based on an EC cell, from 4.0–6.0 mg/L in raw water to lower than 0.5 mg/L. Further, EC was applied to treat groundwater from Shivdaspura (Rajasthan): Sinha et al. [43], reported initial F levels of 5.0 mg/L were reduced to 0.2 mg/L plus. Then, Emamjomeh and Sivakumar [40] confirmed the technique as an effective electro protocol for domestic and industrial water defluoridation.

On their part, Vasudevan et al. [44] compared the performance of different electrodes in an EC protocol and found that F removal efficiency reached 96% with a magnesium alloy anode and a stainless steel cathode at a current density of 0.2 A/dm^2 and pH of 7.0 more recently Khan et al. [45], confirmed earlier findings by Takdastan et al. [46] that aluminum electrodes were more efficient and economical in F removal than iron electrodes. In related analyzes, Hu et al. [47] showed that the efficacy of an EC system in water defluoridation was controlled by the molar ratio of hydroxide and F to Al(III) and related that optimum activity coefficients for defluoridation in coagulation and electrocoagulation are both close to 3.

3. Membrane methods

Membrane methods are those that employ the use of a casing that selectively separates a component in water. They include electrodialysis, reverse osmosis, and nanofiltration [48].

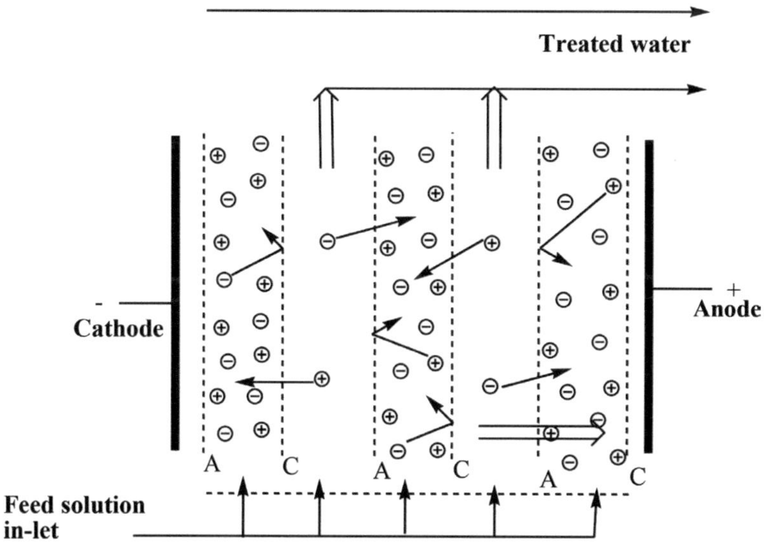

Figure 2.
Schematic representation of an electrodialysis unit.

3.1 Electrodialysis

Electrodialysis (ED) technique uses an electric field to separate ions of one charge from the counter ions [49]. A typical ED unit (**Figure 2**) consists of about 400 alternating cation- and anion-exchange membranes, which are 0.5–2.0 mm wide, sandwiched between an anode and a cathode in a cell [25].

The membranes have charged groups bound into polymeric substrates which attract and adsorb mobile counter ions. The anionic-exchange membranes permeate cations only, while the cation-exchange membranes permeate anions but trap the cations. Under an electric field, cations and anions move in opposite directions and the membranes capture respective ions resulting in alternating cells of ion-concentrated solutions called concentrates and ion-depleted solutions referred to as dilutes [50]. ED is an efficient defluoridation protocol and the sludge volume generated is low. However, the overall protocol is costly, complex, and requires reliable source power and specialized personnel to operate. The process is also non-selective and removes essential ions required for quality drinking water [51].

The technique has been applied to defluoridation of saline water with 3000 mg/L total dissolved salts (TDS) and 3.0 mg/L F [52]. Elazhar et al. [53] compared the performance of ED and nanofiltration (NF). Kabay et al. [54], on the other hand, was able to optimize a water defluoridation process and evaluated its mass transfer and energy use efficiency. ED was applied in Brazil with 97% defluoridation efficiency [49]. In India, ED was applied to saline water with high TDS of 5000 mg/L and 10 mg/L F levels [51] and it has been reported that ED was used to treat brackish water with 2.9 mg/L to just 0.4 mg/L [F] [55].

3.2 Reverse osmosis

Reverse osmosis (RO) is a membrane process in which dissolved pollutants are removed by applying pressure on raw water to force it through a semi-permeable membrane against the osmotic pressure (**Figure 3**) [56].

The level and rate of contaminants removal depend on the sizes and electrical charge of the polluting ions [57]. It is found that RO is efficient and generates little sludge, but it is expensive to install and to run—it requires specialized personnel and reliable electric power to generate necessary pressures [48]. Some of these limitations of RO can, however, be bypassed. A study in Tanzania [58], for example, applied nanofiltration (NF) and reverse osmosis (RO), with an autonomous membrane system, which was powered by solar energy and the tested membranes could achieve the WHO drinking water standards [59]. The process reached 1000–2500 L daily total permeate volume of portable water with an additional 3500–5000 L of non-potable water fit irrigation and washing. However, the integration of such advanced

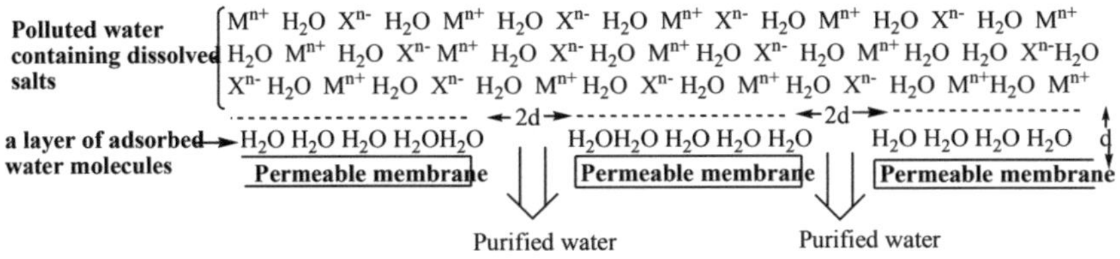

Figure 3.
Schematic representation of reverse osmosis.

technologies has not proven successful in many rural areas of developing nations where the necessary power is always available.

3.3 Nanofiltration

Nanofiltration (NF) operates on the same principle as reverse osmosis but the membranes have larger pores [48] offering less resistance to the flow of solvent and solute particles. The procedure is, therefore, able to operate at much lower pressures reducing energy costs. The retention of solutes is ascribed to steric and charge effects and the procedure is selective and considered to be suited to defluoridation of brackish waters. RO/NF has been applied to the treatment of various groundwater contaminants in India [60] and in the efficient removal of F and salinity from high-F brackish water at a village scale in Senegal [61]. In Finland, Kymenlaakso Water Limited, which is a public company, has operated a 6000 m^3/day water RO plant with a permeate [F] of <0.03 mg/L since 2003 [62]. Richards et al. [63] evaluated the effect of speciation on the retention of F by NF and RO and found that F retention was independent of pH. In a study realized in Tunisia, F removal from water and from wastewater by NF was found to be controlled by trans-membrane pressure, feed water concentration, ionic strength, type of counter-ions, and pH and higher retentions were linked to pH values and vice versa [64]. Some researchers have reported F retention efficiencies of an NF process of 70% [65] mark above pH 7 but another team of researchers in France reported an RO F rejection efficiency greater than 98% [66].

Nonetheless, the protocol continues to attract the interest of researchers from around the globe [67]. Furthermore, hybrid treatments with sequential use of two or more simple techniques have become common in the recent past. For instance, filtration and ultrafiltration as subsequent treatment of coagulation have been recently tested for water defluoridation by a team of workers in India [68].

4. Distillation

4.1 Solar distillation

Distillation is a physical procedure in which water is converted to steam, and then the steam is condensed back into liquid water. The dissolved salts remain in the brackish water that is left behind. Solar distillation takes advantage of abundant sun rays and is the most applicable technique to circumvent the high costs of electricity. The method is simple, clean, and effective but the resulting wastewater must be removed to prevent encrustation of the vessels and be disposed of with care due to their high salinity. Solar distillation can be used from household scale to large industrial scale. Otherwise, it is inexpensive to run but the initial installation costs are big.

Solar distillation units have been used world-over to treat brackish water [69]. A pilot project of Solar driven membrane distillation has been operated on a small village level at Robanda in Tanzania [70]. Also, a similar solar water defluoridation unit has been built using local materials and successfully operated in Bongo District, Ghana, [71].

4.2 Membrane distillation

Membrane distillation uses a hydrophobic membrane with air-filled pores. The surface tension of the feed water and distillate prevents the water from entering the membrane pores keeping it out of the membrane. Water vapor pressure difference is then generated by applying sufficient temperature difference across the membrane. This is accomplished by heating the feed water and cooling the distillate at the other side of the membrane to cause a flow of water vapor through the membrane and result in distillate condensation [70].

Naidu et al. [71] evaluated the applicability of a modified design vacuum enhanced-multi-effect membrane distillation for drinking water and projected a 70% recovery ratio for a scaled-up unit. The feasibility of a direct contact membrane distillation (DCMD) process to recover F contaminated waters was also tested and up to 99% rejection of F was reported [72]. Boubakri et al. [73], using a similar DCMD process based on polyvinylidene F membrane, observed high thermal efficiency and high permeate flux favored by elevated temperatures.

5. Adsorption

In adsorption, raw water is passed through a bed containing a material that is able to retain F by physical, chemical, or ion exchange mechanisms. Adsorption of F onto solid adsorbents occurs through [74]: external mass transfer; surface adsorption; and, intra-particle diffusion processes. The technique has gained popularity in water defluoridation because it offers satisfactory results; it is simple, affordable, and eco-friendly. Many adsorbents, including alumina, clays, polymeric ion-exchange resin, activated carbons, biosorbents, and layered double hydroxides have been studied for water defluoridation [48].

5.1 Activated alumina

Activated alumina, Al_2O_3, is dehydrated $Al(OH)_3$, which is prepared by heating $Al(OH)_3$ at 300–600°C. It was first used in water defluoridation in the US in 1952, later in many other countries including China, Thailand, India [75], South Africa, and Ethiopia [76] by the 1980s. It is now widely used in many other countries of the world [77]. Alumina has one of the highest water defluoridation efficacies. In one study, aluminum hydroxide, which is a form of hydrated alumina was reported with an exceptional F adsorption capacity of 116.75 mg/g [78]. Recently hydrated alumina modified NaA zeolite was reported to have high F adsorption capacities of 104 mg/g [79]. The use of alumina in water defluoridation is, thus, widespread [39] but it is costly, requires frequent adsorbent regeneration, and the adsorbent gets fouled easily from the dissolved solids in the water [25].

5.2 Clay adsorbents

Soils and clays present high prospects of application in water defluoridation. This is mainly because they are almost always: (1) available in natural abundance; (2) stable and usable in different water conditions; (3) have high adsorption capacities; (4) easy to prepare; and, (5) are eco-friendly [80]. The specific reactions F adsorptions at the soil surfaces are heterogeneous and the particular choice of soil adsorbent for water defluoridation is controlled by its known adsorption capacities, availability, and the desired physicochemical properties. Consequently, soils are among the most studied matrices for water defluoridation.

Nonetheless, minerals, which have attracted the highest attention for water defluoridation research include: apatite, calcareous minerals, diatomite, attapulgites, and ferric minerals. The apatite minerals because are known to control the natural exchange of F in soil-water solutions in the environment [81]. Fan et al. [82] evaluated the capacity of hydroxyapatite, fluorspar, calcite, and quartz for water defluoridation. They found that F adsorption capacities for the minerals decreased from the apatite to quartz thus: hydroxyapatite > fluorspar > quartz activated using ferric ions > calcite > quartz. o, many workers have studied the capacity of apatite to enhance limestone, for example, and reported a maximum F adsorption capacity of 3.83 mg/g [83]. Else, it is often found that many calcareous minerals exhibit limited adsorption capacities for F [84]. In a study conducted by Kumar and Gupta [85], the authors investigated fluoride adsorption onto activated diatomite and found that the maximum defluoridation capacity of the mineral was 71.97 mg/kg. Other researchers have, however, reported a more enhanced defluoridation capacity of 51.1 mg/g for the [86]. These have also been collaborated most recently by Taabu et al. [87].

The adsorption of F onto modified attapulgite has been studied widely [88, 89]. The F adsorption capacity for the mineral approximates 24.55 mg/g. Hamdi and Srasra [90] found that water defluoridation capacities for some Tunisian soils was 55.8071.94 mg/g. However, other soils including ferrihydrite and kaolinite-ferrihydrite associate [91], ferric polymineral [92], lateritic minerals [93], clays [94], zeolites [95] and siliceous minerals [96] have been evaluated. Clearly, the capacity of clays to sorb F is greatly varied between the minerals and it is controlled mainly by their mineralogy and the operative conditions [80].

5.3 Ion-exchange resins

Defluoridation protocols based on the ion-exchange technique use charged anion resins that substitute anions in the substrate structure (normally chlorides) for F ions in the water [97]. The resin exchange sites are made of adsorbed cations (usually calcium) [98]. The natural polymeric organic resins, chitin/cellulose composites, are among the adsorbents with the greatest potential for water defluoridation. The use of natural polymeric materials has additional advantages because they are readily available in nature. Subsequently, natural polymers have been studied with varying adsorption efficiencies such as for chitosan (8.10 mg/g) [99], nanocellulose/polyvinyl alcohol composite, agglutinin derived from *Strychnos potatorum L.* seed (11.363 mg/g) [100] and chitosan-zirconia-ferrosoferric oxide composites (17.81 mg/g) [101]. Much higher adsorption capacities of 45.45 mg/g and 52.63 mg/g have, however, been reported for gamma degraded chitosan-Fe(III) beads [102] and for zirconia modified chitosan beads [103], respectively. Furthermore, a sorption potential of 48.78 mg/g has been reported for β-cyclodextrin grafted upon nanoscale titania surfaces [104].

The main challenge of the use of natural polymers in water purification is their liability to chemical and biological degradation. Also, the F ions tend to bind irreversibly into the exchange sites of the resins degrading the membranes. Then it is found that F removal using ion-exchange resins is often limited by low ionic selectivity [105]. Plus, commercial resins are expensive and require continuous regeneration and the spent adsorbents are non-biodegradable, they persist in the environment and must be disposed of very carefully.

5.4 Metal: organic frameworks

There is an emerging class of F adsorbents in fabricated metal–organic frameworks. This are basically multivalent metal ions intercalated in large organic molecules to enhance the synergic F uptake by the metal ions and the organic substrates. Jeyaseelan et al. [106] investigated defluoridation capacities of three MOF's including fumaric acid–based metal–organic frameworks (MOFs) using Zr^{4+}, La^{3+}, and Fe^{3+} metal ions and found that they had comparable defluoridation capacities of 4.920, 4.925, and 4.845 mg/g, respectively.

5.5 Carbonaceous adsorbents

Carbonaceous F adsorbents include activated carbons from plants and animal biomasses, carbonaceous mineral adsorbents, and graphene.

5.5.1 Plant biomass-derived carbonaceous F adsorbents

Many activated carbons have been studied for water defluoridation. Hanumantharao et al. [107], for example, evaluated *Acacia farnesiana* carbon and reported a low defluoridation capacity of 0.268 mg/g. Similar limited water defluoridation capacities (< 1.5 mg/g) have also been reported for carbons of: *Neem* [108], *Tamarindus indica* fruit shells [109], family fruit [110], zirconium-impregnated coconut shell [111], rice straw [112], zirconium impregnated cashew nutshell [113], pine cone [114], and zirconium impregnated coconut fiber [115]. In contrast, studies by Mondal et al. [116] using sugarcane charcoal revealed F uptake capacities of 7.33 mg/g. Similarly, investigations using *Pithacelobium dulce*, *Ipomoea batatas*, and *Peltophorum ferrugineum* carbons showed defluoridation capacities of 78.96, 76.62, and 74.48, respectively [117] but much higher defluoridation capacities of 142.86 mg/g and 230.61 meq/g have been reported for *Delonix regia* pod carbon [118], certain carbon nanostructures [119] and for activated coffee husks carbons [55], respectively. In general water defluoridation using activated carbons has been shown to be pH-dependent and most carbon adsorbents have the highest F removal at acidic pH < 3 values [80]. Plus, the adsorbent particle appears to play a leading role in controlling the adsorption efficiency – high sorption occurs for the lowest size.

5.5.2 Borne char and activated animal charcoal

Bone char is the oldest known water defluoridation agent, and it was first used in USA from 1940s to the 1960s. The technique was later introduced into other countries and it is now among the most used methods in the developing countries of the world. The method involves the use of animal charcoal which is packed into columns and water percolated through the charcoal media [120]. Mutheki et al. [121] compared to the field and laboratory performance of bone char filters and other filters based on a combination of bone char and calcium-phosphate pellets. They found average uptake F capacities to be 1.2 ± 0.3 mg/g and 3.0 ± 1.0 mg/g, respectively. A study to explore activated carbon from fish bladder showed a maximum F removal of 1.43 mg/g [122]. In related work, Kawasaki et al. [123] who investigated four types of animal biomass and Singanan [124] who studied certain bone char adsorbents for F removal, reported more or less similar defluoridation potential. However, some authors reported 99% F removal from borehole water containing 11 mg/L F based on a cartridge bone char affixed onto a domestic faucet as a flow-through defluoridizer [125]. Therefore, some authors contend that F removal bone char is an efficient protocol for defluoridation of brackish water [22]. However, care must be taken about the preparation of the bone char, related to the taste and odor released into the treated water [126]. Besides, large amounts of organic materials are needed for gasification by expensive treatment at 800–1400 K in an inert atmosphere to obtain the adsorbents.

5.5.3 Graphene

Graphene is an emerging carbon material with sp^2-hybridized single-carbon atom-layer structure [127] that is also a promising adsorbent for water defluoridation [128]. Some workers have studied amine grafted graphene oxide encapsulated chitosan hybrid beads for water defluoridation and found a defluoridation capacity of 4.65 mg/g [129]. However, Li et al. [127], while studying graphene samples obtained from exfoliating graphite materials, reported high F adsorption capacities of 17.65 mg/g. Also, a team investigating the F adsorption by graphene-aluminum-silver-carbon quantum dots reported an adsorption capacity of 12.04 mg/g [130]. Nonetheless, some workers have shown that enhancing graphene oxide using cupric oxide, improved its F uptake capacity to 34 mg/g [131]. The results collaborate with those of a team of researchers, which used aluminum modified graphene oxide (GO) and showed it to have superior F removal efficacies of 38.31 mg/g [132]. This shows the high potential use of graphene in F water remediation.

Besides activated carbons and graphene, the application of carbonaceous minerals to water defluoridation has been reviewed elsewhere [80]. However, Abe et al. [133] reported that the water defluoridation capacities of various carbons follow the order: bone char > coal charcoal > wood charcoal > carbon black > petroleum coke.

5.6 Biosorption

Biosorption utilizes animal and plant remains in the pulverized form "as is" without prior gasification or charring. A study realized in Tanzania, reported F removal efficiencies, which was 4.1–47.3%, for several biosorbents [134]. In a particular defluoridation study, which was conducted by Yadav et al. [135] using three agricultural-based biomasses as adsorbents tested on groundwater containing 5 mg/L F, the authors reported F removal efficacies of 40–58%. Elsewhere, Gandhi and Sekhar [136], found that the F biosorption capacity of *Strychnos potatorum* seed powder was 0.9945–1.052 mg/g. Further recently, the F removal of palm kernel shell-based adsorbent was evaluated and found to be 2.35 mg/g [137]. It is found, therefore, that, plant biomasses exhibit limited F uptake capacities when compared to other adsorbents unless they are formidably treated.

Also, defluoridation studies utilizing algal biosorbents derived from *Spirogyra IO1* [138], *Ulva fasciata* sp. [139], *polyalthia longifolia* [140], and *Spirogyra* IO2 to adsorb F [141] have been reported. In the defluoridation evaluation of *Spirogyra* IO2, [141], for example, sorption capacity of 1.272 mg/g was reported. Also, the defluoridation capacities of fungal biosorbents have been studied. In evaluating *Fusarium oxysporum* to remove F from water, low defluoridation capacity of just 19% was reported [142]. These similar findings were collaborated with those by Ramanaiah et al. [143] fungal biosorbent of *Pleurotus ostreatus* 1804. However, other researchers have shown that the F biosorption potential of *Saccharomyces* sp. biomass reached 91% of F removal [144]. Similarly, *Aspergillus* and Calcium treated *Aspergillus* biosorbents revealed

F adsorption capacities of 8.09 mg/g [145]. Differences in the capacity of various biomasses to sequestrate F from water are related to, among other factors, differences in active functional groups in the biomasses [134].

Thus, even though Mukkanti and Tembhurkar [146] have recently reported a high F adsorption capacity of 26.31 mg/g for an adsorbent developed from clamshell waste, it is apparent that the usual water defluoridation capacities of untreated biomasses are low when compared to the other adsorbents [90, 118]. Furthermore, the source biota for the prized F biosorbents may be non-existent in the regions where they are needed for easy defluoridation of water. Plus, untreated biomasses degrade easily under chemical and biological attacks.

5.7 Layer double hydroxides

Layered double hydroxides (LDHs), also called anionic clay and hydrotalcite-like compounds, are a "host-guest" layered materials, which have the general formula $[M^{2+}{}_{1-x}M^{3+}{}_{x}(OH)_2]^{x+}(A^{n-}{}_{x/n}).mH_2O$, where M^{2+} and M^{3+} are metal cations that occupy octahedral positions in hydroxide layers; x is the molar ratio $M^{3+}/(M^{2+} + M^{3+})$ and A denotes interlayer charge-compensating anions [147]. LDHs have attracted a lot of attention as F adsorbents in the recent past. Lu et al. [148] assessed F removal based on NiAl layered double hydroxides (NiAl-LDHs) and reported a low equilibrium F concentration of just 0.2388 mg/L in the treated water. Sadik et al. [149] also reported high F removal rates of 99.2% for calcined LDHs synthesized from seawater (LDHsw). New data have provided the maximum F adsorption capacity of 6.67 mg/g for $Fe_3O_4/Al(OH)_3$ [150] and 12.63 mg/g for tri-metal Mg/Ce/Mn oxide-modified diatomaceous matrix [151]. However, the mechanism of F adsorption onto an LDH and calcined layered double hydroxide (CLDH) had been earlier evaluated with maximum defluoridation capacities of 1.3 mg/g and 20 mg/g, respectively [152]. The F sorption quantities were somehow similar to the 22.78 mg/g and 20.28 mg/g that have recently been reported for Ce-Ti and Ce-Ti/Fe3O4 hybrid oxides, respectively [153].

However, several studies have reported enhanced defluoridation capacities for LDHs. In a study involving calcined Mg–Al–CO_3 LDHs, for example, competitive F adsorption was evaluated and water defluoridation capacity of 1.94 mmol/g was reported [154]. Then, Kang et al. [155] reported water defluoridation capacity of 50.91 mg/g for Mg/Fe l CLDHs. Furthermore, other studies have documented a high F adsorption capacity of 146.6 mg/g for Ca-Al LDHs [156] and 270.3 mg/g for Fe–Mg–La triple-metal hydroxide composite [157]. Consequently, other authors have focused on the optimization of defluoridation conditions for LDHs. Elhalil et al. [158], as such, while evaluating showed that optimum adsorbent dosages of water defluoridation using calcined Mg/Al LDH were in the range of 0.29–0.8 g/L. Also, the suggested F adsorption equilibrium time [149, 156] and solution acidity [149, 155] of LDHs are within 1 h and pH 6–7, respectively.

6. Phytoremediation

This is a technique of defluoridation and removal of other contaminants from the environment, which uses plants to absorb and accumulate excess F from soil and water through their roots into their systems. The plants are then removed at the predetermined time and disposed of safely. Several researchers evaluated technology phytoremediation with varying degrees of success. The tolerance capacity of *Solanum tuberosum* to accumulate F, has been tested and found that after 87 days, the F levels in the leaves, root, shoot, and potato tuber of the plants had increased to 3.96, 3.02, 2.8, and 1.56 mg, respectively [159]. Some researchers also tested the uptake of Al and F by four green algae species and found that *Pseudokirchneriella subcapitata* showed the highest aluminum and fluoride absorption under the test conditions [160]. Sirisha et al. [161] studied phytoremediation of Cr and F in industrial wastewater using the aquatic plant *ipomoea aquatica plant*. They found that the F removal rapidly reached 37% in just 10 min and similar results have been found by researchers in related tests [162] indicating that phytoremediation id a promising green technology for use in environmental fluoride remediation [163].

7. Conclusions

The current work, which was based on a systematic collection of literature data aimed at providing, in a concise and precise manner, an update on the techniques employed to combat the detrimental effects of human overexposure to F so as to focus the attention of stakeholders to the direction of science in the field of F mitigation in the world. From the foregoing discussions covered in this paper, the following conclusions are made:

i. Water defluoridation techniques applied to the world-over can be classified broadly as chemical, physical, membrane-based, or adsorptive.

ii. The different defluoridation protocols differ in applicability and feasibility, depending on the desired levels of water defluoridation and resource availability.

iii. Adsorptive methods appear to present greater prospects in water defluoridatiore because they are simple, efficient, and cost-effective.

iv. LDH's and soil adsorbents are the most studied adsorbents for water defluoridation, but the latter has a competitive edge over the former as they are readily available, easier to prepare and use, and more environmentally convenient.

v. The efficacy of other adsorbents, such as activated carbons, is greatly controlled by the mode of adsorbent preparation and the adsorbent particle sizes.

Acknowledgements

This publication is part of the FLOWERED project that received funding from the EU's Horizon 2020 research and innovation program under grant agreement No 690378. It is also in fond memories of Prof. Giorgio Ghiglieri, Department of Chemical and Geological Sciences, University of Cagliari, Sardinia, Italy, who was the Coordinator for the FLOWERED Project.

Author details

Enos Wamalwa Wambu[1*], Franco Frau[2], Revocatus Machunda[3], Lilliane Pasape[4], Stephen S. Barasa[1] and Giorgio Ghiglieri[2†]

1 Department of Chemistry and Biochemistry, University of Eldoret, Eldoret, Kenya

2 Department of Chemical and Geological Sciences, University of Cagliari, Sardinia, Italy

3 Department of Water and Environmental Sciences and Engineering, Nelson Mandela African Institution of Science and Technology, Arusha, Tanzania

4 Department of Innovation, Technology Management an Entrepreneurship, School of Business Studies and Humanities, Nelson Mandela Institution of Science and Technology, Arusha, Tanzania

*Address all correspondence to: wambuenos@yahoo.com

† Deceased.

References

[1] Dean HT, Elvove E. Some epidemiological aspects of chronic endemic dental fluorosis. American Journal of Public Health and the Nation's Health. 1936;**26**:567-575

[2] Waugh DT, Potter W, Limeback H, Godfrey M. Risk assessment of fluoride intake from tea in the Republic of Ireland and its implications for public health and water fluoridation. International Journal of Environmental Research and Public Health. 2016;**13**:259-280

[3] Levy SM. An update on fluorides and fluorosis. Journal of the Canadian Dental Association. 2003;**69**:286-291

[4] Ayoob S, Gupta K. Fluoride in drinking water: A review on the status and stress effects. Critical Reviews in Environmental Science and Technology. 2006;**36**:433-487

[5] Mascarenhas AK. Risk factors for dental fluorosis : A review of the recent literature. Pediatric Dentistry. 2000;**269**(22):269-277

[6] Browne D, Whelton H, O'Mullane D. Fluoride metabolism and fluorosis. Journal of Dentistry. 2005;**33**:177-186

[7] Li Y, Liang C, Slemenda CW, Ji R, Sun S, Cao J, et al. Effect of long-term exposure to fluoride in drinking water on risks of bone fractures. Journal of Bone and Mineral Research. 2001;**16**:932-939

[8] UNICEF, WHO, Progress on sanitation and drinking water. 2010 Update, 2010. Available from: http://www.unwater.org/downloads/JMP_report_2010.pdf

[9] Susheela A. Fluorosis in developing countries: Remedial measures and approaches. Proceedings-Indian National Science Academy Part B. 2002;**68**:389-400

[10] Manji F, Baelum V, Fejerskov O. Dental fluorosis in an area of Kenya with 2 ppm fluoride in the drinking water. Journal of Dental Research. 1986;**65**:659-662

[11] Erdal S, Buchanan SN. A quantitative look at fluorosis, fluoride exposure, and intake in children using a health risk assessment approach. Environmental Health Perspectives. 2005;**113**:111-117

[12] Jacintha TGA, Rawat KS, Mishra A, Singh SK. Hydrogeochemical characterization of groundwater of peninsular Indian region using multivariate statistical techniques. Applied Water Science. 2017;7:3001-3013

[13] Ghiglieri G, Balia R, Oggiano G, Pittalis D. Prospecting for safe (low fluoride) groundwater in the eastern African rift : The Arumeru District (northern Tanzania). Hydrology and Earth System Sciences. 2010;**14**:1081-1091

[14] Gumbo FJ, Mkongo G. Water Defluoridation for rural fluoride affected communities in Tanzania. In: Spittle B, editor. 1 St Work. Fluorosis Defluoridation Held Ngurdoto, Tanzania, October 18-22, 1995. Dunedin, New Zealand: The International Society for Fluoride Research; 1995. pp. 109-114

[15] Kanduti D, Sterbenk P, Artnik B. Fluoride: A review if use and effects on health. Mater Sociomed. 2016;**28**:133-137. DOI: 10.5455/msm.2016.28.133-137

[16] Turner BD, Binning P, Stipp BDS. Fluoride removal by calcite: Evidence

for fluorite precipitation and surface adsorption. Environmental Science & Technology. 2005;**29**:9561-9568

[17] Hendrickson K, Vik EA. Adsorption in Water Treatment: Fluoride Removal. Oslo, Norway: Norwegian Institute for Water Research; 1984

[18] Tebbutt THY. Principles of Water Quality Control. 5th ed. Amsterdam: Butterworht-Heinemann; 1998

[19] Federal-Provincial Committee on Drinking Water. Guidelines for Canadian Drinking Water Quality: Guideline Technical Document-Fluoride. Ottawa: Canada; 2010

[20] Opinya GN, Pameijer CH. Simple defluoridation procedures for Kenyan borehole water. Community Dentistry and Oral Epidemiology. 1987;**15**:60-62

[21] Obasi Onuoha U. Evaluation of Alumina Sorptiom System for Removal of Fluoride from Water. Doctorate Dissertation, Texas Tech University; 1983

[22] Renuka P, Pushpanjali K. Review on Defluoridation techniques of water. International Journal of Engineering Science. 2013;**2**:86-94

[23] Korir H, Mueller K, Korir L, Kubai J, Wanja E, Wanjiku N, et al. The development of bone char—based filters for the removal of fluoride from drinking water. In: 34th WEDC International Conference, Addis Ababa, Ethiopia, 2009 Water, Sanitation and Hygiene: Sustainable Development and Multisectoral Approaches. Addis Ababa, Ethiopia; 2009. pp. 1-6

[24] Dahi E. Contact precipitation for defluoridation of water. In: 22nd WEDC Conference Reaching the Unreached Challenges for 21st Century. New Delhi, India; 1996. pp. 262-265

[25] Consultative Committee of the Kenya Bureau of Standards (KEBS), Excessive Fluoride in Water in. Kenya, Nairobi, Kenya; 2010

[26] Hu CY, Lo SL, Kuan WH. Effects of the molar ratio of hydroxide and fluoride to Al(III) on fluoride removal by coagulation and electrocoagulation. Journal of Colloid and Interface Science. 2005;**283**:472-476

[27] Bolto B, Gregory J. Organic polyelectrolytes in water treatment. Water Research. 2007;**41**:2301-2324

[28] Aoudj S, Drouiche N, Hecini M, Ouslimane T, Palaouane B. Coagulation as a post-treatment method for the Defluoridation of photovoltaic cell manufacturing wastewater. Procedia Engineering. 2012;**33**:111-120

[29] Stehouwer M. Defluoridation and Natural Organic Matter Removal in Drinking Waters by Alum Coagulation. Doctoral Dissertation, The University of Texas; 2014

[30] Deng L, Wang Y, Zhou J, Huang T, Sun X. Impact of acid-base conditions on de fl uoridation by induced crystallization. Journal of Industrial and Engineering Chemistry. 2019;**83**:35-45

[31] Ingallinella AM, Pacini VA, Fernández RG, Vidoni RM, Sanguinetti G. Simultaneous removal of arsenic and fluoride from groundwater by coagulation-adsorption with polyaluminum chloride. Journal of Environmental Science and Health. Part A, Toxic/Hazardous Substances & Environmental Engineering. 2011;**46**:1288-1296

[32] Linzhi Z, Yihan SUN, Cheng H. Research on coagulation/sedimentation process for simulation of fluorine-containing wastewater treatment.

Applied Mechanics and Materials. 2013;**361-363**:755-759

[33] Solanki YS, Agarwal M, Maheshwari K, Gupta S. Removal of fluoride from water by using a coagulant (inorganic polymeric coagulant). Environmental Science and Pollution Research. 2020;**1**:3897-3905

[34] Wang Z, Wang Z, Su J, Ali A, Zhang R, Yang W, et al. Synergistic removal of fluoride from groundwater by seed crystals and bacteria based on microbially induced calcium precipitation. Science of The Total Environment. 2021;**806**:1-9

[35] Bulusu KR, Nawlakhe WG. Defluoridation of water with activated alumina: Batch operations. Indian Journal of Environmental Health. 1988;**30**:262-299

[36] Nawlakhe WG, Paramasivam R. Defluoridation of potable water by Nagonda technique. Current Science. 1993;**65**:743-748

[37] Agarwal KC, Gupta SK, Gupta AB. Development of nwe low-cost defluoridation technology. Water Science and Technology. 1999;**40**:167-173

[38] Shankar R, Singh L, Mondal P, Chand S. Removal of lignin from wastewater through electro-coagulation, world. Journal of Environmental Engineering. 2013;**1**:16-20

[39] Rao N. Fluoride and environment- a review. In: Bunch MJ, Suresh VM, Kumaran TV, editors. Proceedings of the Third International Conference Environmental Health Chennai, India, 15-17 December, 2003. York University, Chennai, India: Department of Geography, University of Madras and Faculty of Environmental Studies; 2003. pp. 386-399

[40] Emamjomeh MM, Sivakumar M. Review of pollutants removed by electrocoagulation and electrocoagulation/flotation processes. Journal of Environmental Management. 2009;**90**:1663-1679

[41] Emamjomeh MM, Sivakumar M, Schafer AI. Fluoride removal by using a batch electrocoagulation reactor. In: 7th Annual Environmental Engineering Research Event (EERE) Conference, Marysville, Victoria, Australia. 2003. P. 143-152

[42] Khatibikamala V, Torabiana A, Janpoora F, Hoshyaripour G. Fluoride removal from industrial wastewater using electrocoagulation and its adsorption kinetics. Journal of Hazardous Materials. 2010;**179**:276-280

[43] Sinha R, Khazanchi I, Mathur S. Fluoride removal by a continuous flow electrocoagulation reactor from groundwater of Shivdaspura. International Journal of Engineering Research and Applications. 2012;**2**:1336-1341

[44] Vasudevan S, Lakshmi J, Sozhan G. Studies on a Mg-Al-Zn alloy as an anode for the removal of fluoride from drinking water in an electrocoagulation process. CLEAN – Soil, Air, Water. 2009;**37**:372-378

[45] Khan SU, Asif M, Alam F, Khan NA, Farooqi IH. Optimizing fluoride removal and energy consumption in a batch reactor using electrocoagulation : A smart treatment technology. In: Ahmed S, Abbas S, Zia H, editors. Smart Cities - Opportumities and Challenges. Singapore: Springer; 2020. pp. 767-778

[46] Takdastan A, Tabar SE, Islam A, Bazafkan MH, Naisi AK. The effect of the electrode in fluoride removal from drinking water by electro coagulation

process. In: International Conference on Chemical and Biological Sciences, March 18-19th, 2015. 2015. pp. 39-44

[47] Hu S, Yan L, Chan T, Jing C. Molecular insights into ternary surface complexation of arsenite and cadmium on TiO_2. Environmental Science & Technology. 2015;**49**:5973-5979

[48] Mohapatra M, Anand S, Mishra BK, Giles DE, Singh P. Review of fluoride removal from drinking water. Journal of Environmental Management. 2009;**91**:67-77

[49] Zeni M, Riveros R, Melo K, Primieri R, Lorenzini S. Study on fluoride reduction in artesian well—Water from electrodialysis process. Desalination. 2005;**185**:241-244

[50] Reuther CG. Saline solutions: The quest for fresh water. Environmental Health Perspectives. 2000;**108**:A78-A80

[51] Adhikary SK, Tipnis UK, Harkare WP, Govindan KP. Defluoridation during desalination of brackish water by Electrodialysis. Desalination. 1989;**71**:301-312

[52] Sahli MA, Annouar A, Tahaikt S, Mountadar M, Soufiane A, Elmidaoui A, et al. Fluoride removal for underground brackish water by adsorption on the natural chitosan and by electrodialysis. Desalination. 2007;**212**:37-45

[53] Elazhar F, Tahaikt M, Zouahri A, Taky M, Hafsi M, Elmidaoui A. Defluoridation of Moroccan groundwater by Nanofiltration and Electrodialysis : Performances and cost comparison. World Applied Sciences Journal. 2013;**22**:844-850

[54] Kabay N, Ara O, Samatya S, Yüksel Ü, Yüksel M. Separation of fluoride from aqueous solution by electrodialysis: Effect of process parameters and other ionic species. Journal of Hazardous Materials. 2008;**153**:107-113

[55] Ali MBS, Hamrouni B, Dhahbi M. Electrodialytic Defluoridation of brackish water: Effect of process parameters and water characteristics. CLEAN – Soil, Air, Water. 2010;**38**: 623-629

[56] RajeshwarK,IbanezJG.Environmental Electrochemistry: Fundamentals and Applications in Pollution Sensors. Amsterdam: Elsevier science & Technology books; 1997

[57] Shen J, Schäfer AI. Removal of fluoride and uranium by nanofiltration and reverse osmosis : A review. Chemosphere. 2014;**117**:679-691

[58] Shen J, Mkongo G, Abbt-Braun G, Ceppi SL, Richards BS, Schäfer AI. Renewable energy powered membrane technology: Fluoride removal in a rural community in northern Tanzania. Separation and Purification Technology. 2015;**149**:349-361

[59] WHO. Guidelines for Drinking-Water Quality, Third Edition, Incorporating First and Second Addenda. Geneva; 1984

[60] Gedam VV, Patil JL, Kagne S, Sirsam RS, Labhasetwar P. Performance evaluation of polyamide reverse osmosis membrane for removal of contaminants in ground water collected from Chandrapur District. Journal of Membrane Science & Technology. 2012;**2**:117-122

[61] Diawara CK, Diop SN, Diallo MA, Farcy M, Deratani A, Corporation P, et al. Performance of Nanofiltration (NF) and low pressure reverse osmosis (LPRO) membranes in the removal of fluorine and salinity from brackish drinking water.

Journal of Water Resource and Protection. 2011;**3**:912-917

[62] Sehn P. Fluoride removal with extra low energy reverse osmosis membranes: Three years of large scale field experience in Finland. Desalination. 2008;**223**:73-84

[63] Richards LA, Richards BS, Rossiter HMA, Schäfer AI. Impact of speciation on fluoride, arsenic and magnesium retention by nanofiltration/ reverse osmosis in remote australian communities. Desalination. 2009;**248**:177-183

[64] Bejaoui I, Mnif A, Hamrouni B. Performance of reverse osmosis and Nanofiltration in the removal of fluoride from model water and metal packaging industrial effluent performance of reverse osmosis and Nanofiltration in the removal of fluoride from model water and metal packaging Industr, Saparation. Science and Technology. 2014;**49**:1-11

[65] Richards LA, Vuachère M, Schäfer AI. Impact of pH on the removal of fluoride, nitrate and boron by nanofiltration/reverse osmosis. Desalination. 2010;**261**:331-337

[66] Ndiaye PI, Moulln P, Dominguez L, Millet JC. Removal of fluoride from electronic industrial effluent by RO membrane separation. Desalination. 2005;**173**:25-32

[67] Grzegorzek M, Majewska-nowak K, Ahmed AE. Science of the Total environment removal of fl uoride from multicomponent water solutions with the use of monovalent selective ion-exchange membranes. Science of The Total Environment. 2020;**722**:137681

[68] Dubey S, Agarwal M, Gupta AB. Experimental evaluation of sand filtration and ultrafiltration as subsequent treatment of coagulation for fluoride removal. Environmental Progress & Sustainable Energy. 2021;**e13790**:1-11

[69] Antwi E, Cudjoe E, Cudjoe J. Use of solar water distiller for treatment of fl uoride-contaminated water : The case of bongo district of Ghana. Desalination. 2011;**278**:333-336

[70] van Tongeren WGJM, Assink WJ, Appelman WAJ. Pilot Test Solar Driven Membrane Distillation for Drinking Water Production at Robanda Tanzania NWASH Project Result 2: Innovative Fluoride Removal Technology. The Netherlands: Zeist; 2016

[71] Naidu G, Jeong S, Choi Y, Jang E, Hwang T, Vigneswaran S. Application of vacuum membrane distillation for small scale drinking water production. Desalination. 2014;**354**:53-61

[72] Yarlagadda S, Gude VG, Camacho LM, Pinappu S, Deng S. Water recovery from As, U, and F contaminated ground waters by direct contact membrane distillation process. Journal of Hazardous Materials. 2011;**192**:1388-1394

[73] Boubakri A, Bouchrit R, Hafiane A. Fluoride removal from aqueous solution by direct contact membrane distillation: Theoretical and experimental studies. Environmental Science & Technology. 2014;**21**:10493-10501

[74] Habuda-Stanić M, Ravančić ME, Flanagan A. A review on adsorption of fluoride from aqueous solution. Materials. 2014;**7**:6317-6366

[75] Dahi E. The state of art of small community Defluoridation of drinking water. In: Dahi E, Rajchagool S, Osiriphan N, editors. 3rd International Workshop on Fluorosis Prevention and Defluoridation of Water, Chiang

Mai, Thailand, November 20-24, 2000. Dunedin, New Zealand: International Society of Fluoride Research; 2000. pp. 141-170

[76] Feenstra L, Vasak L, Griffioen J. Fluoride in Groundwater: Overview and Evaluation of Removal Methods. International Groundwater Resources Assessment Centre Utrecht, Utrecht, The Netherlands; 2007. Available from: www.igrac.nl

[77] Venkobachar C, Iyengar L, Mudgal A. Household defluoridation of drinking water using activated alumina technology, Nazreth, Ethiopia November 19-25th, 1997. In: Dahi E, Nielsen JM, editors. Proceedings of 2nd International Work Fluorosis Defluoridation Water. Dunedin, New Zealand: International Society for Fluoride Research; 1997. pp. 138-145

[78] Sequeira A, Solache-Ríos M, Balderas-Hernández P. Modification effects of hematite with Aluminum hydroxide on the removal of fluoride ions from water. Water, Air, & Soil Pollution. 2011;**223**:319-327

[79] Naskar MK. Preparation of colloidal hydrated alumina modified NaA zeolite derived from rice husk ash for effective removal of fluoride ions from water medium. Journal of Asian Ceramic Societies. 2020;**8**:437-447

[80] Wambu EWEW, Ambusso WOW, Onindo CO, Muthakia GKGK, Wambu EWEW. Review of fluoride removal from water by adsorption using soil adsorbents -an evaluation of the status. Journal of Water Reuse and Desalination. 2016;**6**:1-29

[81] Willard RL, Campell TJ, Rapp RG. Encyclopaedia of Minerals. 2nd ed. New York: Van Nostrand Reinhold Company; 1990

[82] Fan X, Parker DJ, Smith MD. Adsorption kinetics of fluoride on low cost materials. Water Research. 2003;**37**:4929-4937

[83] Mohan R, Dutta RK. Continuous fixed - bed column assessment for defluoridation of water using HAp - coated - limestone. Journal of Environmental Chemical Engineering. 2020;**8**(4):7-8

[84] Yapo NZS, Gouessé B, Briton H, Aw S, Reinert L, Drogui P, et al. Toxic/hazardous substances and environmental engineering bivalve shells (Corbula trigona) as a new adsorbent for the defluoridation of groundwater by. Journal of Environmental Science and Health, Part A. 2021;**56**:694-704

[85] Kumar P, Gupta P. Defluoridation of domestic waste water by using activated diatomaceous earth in fixed mattress column adsorption system. Oriental Journal of Chemistry. 2021;**37**:594-601

[86] Wambu EW, Onindo CO, Ambusso WJ, Muthakia GK. Fluoride adsorption onto acid-treated diatomaceous mineral from Kenya. Materials Sciences and Applications. 2011;**2**:1654-1660

[87] Taabu SM, Wanjala NF, Bernard A, Godwin M, Alex O, Gloria M. Use of organic binders to enhance Defluoridation and pathogen removal efficiency of diatomaceous earth-based ceramic filters. Africa Journal of Physical Sciences. 2021;**6**:57-64

[88] Zhang J, Xie S, Ho Y. Removal of fluoride ions from aqueous solution using modified attapulgite as adsorbent. Journal of Hazardous Materials. 2009;**165**:218-222

[89] Zhang G, He Z, Xu W. A low-cost and high efficient

zirconium- modified-Na-attapulgite adsorbent for fluoride removal from aqueous solutions. Chemical Engineering Journal. 2012;**183**:315-324

[90] Hamdi N, Srasra E. Retention of fluoride from industrial acidic wastewater and NaF solution by three tunisian clayey soils. Fluoride. 2009;**42**:39-45

[91] Wei S, Xiang W. Surface properties and adsorption characteristics for fluoride of kaolinite, ferrihydrite and kaolinite-ferrihydrite association. Journal of Food, Agriculture and Environment. 2012;**10**:923-929

[92] Wambu EW, Onindo CO, Ambusso W, Gerald K. Equilibrium studies of fluoride adsorption onto a ferric poly – mineral from Kenya. Journal of Applied Sciences and Environmental Management. 2012;**16**:69-74

[93] Wambu EW, Onindo CO, Ambusso WJ, Muthakia GK, Box PO. Fluoride adsorption onto an acid treated lateritic mineral from Kenya: Equilibrium studies, African. Journal of Environmental Science and Technology. 2012;**6**:160-169

[94] Obijole O, Wilson G, Mudzielwana R, Ndungu P, Samie A, Babatunde A. Hydrothermally treated aluminosilicate clay (HTAC) for remediation of fluoride and pathogens from water : Adsorbent characterization and adsorption modelling. Water Resources and Industry. 2021;**25**:100144

[95] Desalegn Y, Melak F, Yitbarek M, Astatkie H. Groundwater for sustainable development Aluminum coated natural zeolite for water defluoridation : A mechanistic insight. Groundwater for Sustainable Development. 2021;**12**:100525

[96] Wambu EW, Onindo CO, Ambusso W, Muthakia GK. Removal of fluoride from aqueous solutions by adsorption using a siliceous mineral of a Kenyan origin. CLEAN – Soil, Air, Water. 2013;**41**:340-348

[97] Alexandratos SD. Ion-exchange resins : A retrospective from industrial and engineering chemistry research. Industrial and Engineering Chemistry Research. 2009;**48**:388-298

[98] Muraviev D, Torrado A, Valiente M. Kinetics of release of calcium and fluoride ions from ion-exchange resins in Artificail saliva. Solvent Extraction and Ion Exchange. 2007;**18**:345-374

[99] Huang R, Yang B, Liu Q, Ding K. Removal of fluoride ions from aqueous solutions using protonated cross-linked chitosan particles. Journal of Fluorine Chemistry. 2012;**141**:29-34

[100] Sowmiya BR, Vasanthy M, Rajakannan V, Ravindran B, Soon WC, Chandrasekaran M, et al. Defluoridation of water with a coagulant, Strychnos potatorum L. seed – Agglutinin. Environmental Technology and Innovation. 2021;**24**:101983

[101] Korde S, Tandekar S, Jugade RM. Journal of environmental chemical engineering novel mesoporous chitosan-zirconia-ferrosoferric oxide as magnetic composite for de fl uoridation of water. Journal of Environmental Chemical Engineering. 2020;**8**:104360

[102] Tandekar S, Saravanan D, Korde S, Jugade R. Materials today : Proceedings gamma degraded chitosan-Fe (III) beads for defluoridation of water. Materials Today: Proceedings. 2020;**24**:1-7

[103] Tandekar S, Saravanan D, Jugade R. Zirconia-chitosan beads

as highly efficient adsorbent for defluoridation of water. Indian Journal of Chemistry. 2020;**59A**:1067-1075

[104] Fallah Z, Isfahani HN, Tajbakhsh M. Removal of fluoride ion from aqueous solutions by titania-grafted β -cyclodextrin nanocomposite. Environmental Science and Pollution Research. 2019;**27**(3):3281-3294

[105] Karthikeyan S, See SW, Balasubramanian R. Simultaneous determination of inorganic anions and selected organic acids in airborne particulate matter by ion chromatography. Analytical Letters. 2007;**40**:793-804

[106] Jeyaseelan A, Naushad M, Viswanathan N. Development of multivalent metal-ion-fabricated Fumaric acid-based development of multivalent metal-ion-fabricated Fumaric acid-based metal – organic frameworks for De fl uoridation of water. Journal of Chemical & Engineering Data. 2020;**65**:2990-3001

[107] Hanumantharao Y, Kishore M, Ravindhranath K. Preparation and development of adsorbent carbon from acacia farnesiana for defluoridation. International Journal of Plant, Animal and Environmental Sciences. 2011;**1**:209-223

[108] Chakrabarty S, Sarma HP. Defluoridation of contaminated drinking water using neem charcoal adsorbent : Kinetics and equilibrium studies. International Journal of ChemTech Research. 2012;**4**:511-516

[109] Sivasankar V, Rajkumar S, Murugesh S, Darchen A. Tamarind (Tamarindus indica) fruit shell carbon: A calcium-rich promising adsorbent for fluoride removal from groundwater. Journal of Hazardous Materials. 2012;**225-226**:164-172

[110] Medikondu K. Potable water defluoridation by lowcost adsorbents from Mimosideae family fruit carbons: A comparative study. International Letters of Chemistry, Physics and Astronomy. 2016;**56**:71-81

[111] Sathish RSS, Raju NSRSR, Raju GSS, Rao GNN, Kumar KA, Janardhana C, et al. Equilibrium and kinetic studies for fluoride adsorption from water on zirconium impregnated coconut Shell carbon fluoride adsorption from water on. Separation Science and Technology. 2007;**42**:769-788

[112] Daifullah AAM, Yakout SM, Elreefy SA. Adsorption of fluoride in aqueous solutions using KMnO4-modified activated carbon derived from steam pyrolysis of rice straw. Journal of Hazardous Materials. 2007;**147**:633-643

[113] Alagumuthu G, Rajan M. Equilibrium and kinetics of adsorption of fluoride onto zirconium impregnated cashew nut shell carbon. Chemical Engineering Journal. 2010;**158**:451-457

[114] Singh R, Surjit T, Katoch S, Modi A. Assessment of pine cone derived activated carbon as an adsorbent in defluoridation. SN Applied Sciences. 2020;**2**:1-12

[115] Janardhana C, Rao GN, Sathish RS, Kumar PS, Kumar VA, Madhav MV. Study on defluoridation of drinking water using zirconium ion impregnated activated charcoals. Indian Journal of Chemical Technology. 2007;**14**:350-354

[116] Mondal NK, Bhaumik R, Roy P, Das B, Datta JK. Investigation on fixed bed column performance of fluoride adsorption by sugarcane charcoal. Journal of Environmental Biology. 2013;**34**:1059-1064

[117] Emmanuel KA, Ramaraju KA, Rambabu G, Rao AV, Autonomous SCRR,

Pradesh A. Removal of fluoride from drinking water with activated carbons prepared from HNO_3 activation - a comparative study. Rasayan Journal of Chemistry. 2008;**11**:802-818

[118] Ajisha MAT, Rajagopal K. Fluoride removal study using pyrolyzed Delonix regia pod, an unconventional adsorbent. International journal of Environmental Science and Technology. 2015;**12**:223-236

[119] Faghihian H, Atarodi H, Kooravand M. Synthesis, treatment, and application of a novel carbon nanostructure for removal of fluoride from aqueous solution. Desalination and Water Treatment. 2014;**0**:1-9

[120] Mjengera H, Mkongo G. Appropriate defluoridation technology for use in fluoritic areas in Tanzania. Physics and Chemistry of the Earth. 2003;**28**:1097-1104

[121] Mutheki PM, Osterwalder L, Kubai J, Korir L, Wanja E, Wambui E, et al. Comparative performance of bone char-based filters for the removal of fluoride from drinking water. In: 35th Water Engineerig and Development Centre (WEDC). International Conference. Loughbrgh, UK. 2011. pp. 1-4

[122] Karuga J, Jande Y, Kim H, King C. Fish swim bladder-derived porous carbon for Defluoridation at potable water pH. Advances in Chemical Engineering and Science. 2016;**6**:500-514

[123] Kawasaki N, Ogata F, Tominaga H, Yamaguchi I. Removal of fluoride ion by bone char produced from animal biomass. Journal of Oleo Science. 2009;**535**:529-535

[124] Singanan M. Defluoridation of drinking water using metal embedded biocarbon technology. International Journal of Environmental Engineering; 2013;**5**(2):150-160

[125] Nyanchaga N, Bailey T. Fluoride contamination in drinking water in the Rift Valley, Kenya and evaluation of the efficiency of a locally manufactured Defluoridation filter. Journal of Civil Engineering, JKUAT. 2003;**8**:79-88

[126] Larsen MJ, Pearce EIF. Defluoridation of drinking water by boiling with Brushite and calcite, caries research;36:341-346. Caries Research. 2002;**36**:341-346

[127] Li Y, Zhang P, Du Q, Peng X, Liu T, Wang Z, et al. Adsorption of fluoride from aqueous solution by graphene. Journal of Colloid and Interface Science. 2011;**363**:348-354

[128] Kumar A, Garima VSK, Share R, Gupta S. Defluoridation studies using graphene oxidebased nanoadsorbents. In: Dehghani ÉH, Karri RR, editors. Green Chemistry and Water Remediation: Research and Applications. Elsevier. 2021. pp. 35-57

[129] Jeyaseelan A, Salem N, Mohammedsaleh K, Katubi M, Naushad M, Viswanathan N. International journal of biological macromolecules design and synthesis of amine grafted graphene oxide encapsulated chitosan hybrid beads for de fl uoridation of water. International Journal of Biological Macromolecules. 2021;**182**:1843-1851

[130] Singh N, Kumari S, Khan S. Improved fluoride removal efficiency using novel defluoridation pencil. Materials Today Communications. 2021;**28**:102521

[131] Mohan S, Basavaiah K. Defluoridation in aqueous solution by a composite of reduced graphene

oxide decorated with cuprous oxide via sonochemical. Arabian Journal of Chemistry. 2020;**13**:7970-7977

[132] Rajput A, Raj SK, Sharma PP, Yadav V, Sarvaia H, Gupta H, et al. Synthesis and characterization of aluminium modified graphene oxide: An approach towards defluoridation of potable water. Journal of Dispersion Science and Technology. 2019;**40**:1101

[133] Abe I, Iwasaki S, Tokimoto T, Kawasaki N, Nakamura T, Tanada S. Adsorption of fluoride ions onto carbonaceous materials. Journal of Colloid and Interface Science. 2004;**275**:35-39

[134] Mwakabona HT, Machunda RL, Njau KN. The influence of stereochemistry of the active compounds on fluoride adsorption efficiency of the plant biomass. American Journal of Chemical Engineering. 2014;**2**:42-47

[135] Yadav AK, Abbassi R, Gupta A, Dadashzadeh M. Removal of fluoride from aqueous solution and groundwater by wheat straw, sawdust and activated bagasse carbon of sugarcane. Ecological Engineering. 2013;**52**:211-218

[136] Gandhi N, Sekhar KBC. Bioremediation of waste water by using Strychnos potatorum seeds (clearing nuts) as bio adsorbent and natural coagulant for removal of fluoride and chromium. Journal of International Academic Research for Multidisciplinary. 2014;**2**:253-272

[137] Bashir M, Salmiaton A, Nourouzi M, Azni I, Harun R. Fluoride removal by chemical modification of palm kernel Shell-based adsorbent: A novel agricultural waste utilization approach. Asian Journal of Microbiology, Biotechnology & Environmental Sciences. 2015;**17**:533-542

[138] Mohan SV, Ramanaiah SV, Rajkumar B, Sarma PN. Biosorption of fluoride from aqueous phase onto algal Spirogyra IO1 and evaluation of adsorption kinetics. Bioresource Technology. 2007;**98**:1006-1011

[139] Kalyani G, Rao GB, Saradhi BV, Kumar YP. Biosorption isotherms of fluoride from aqueous solution on *Ulva Fasciata* SP.-a waste material. International Journal of Applied Environmental Sciences. 2009;**4**:173-182

[140] Bharali RK, Bhattacharyya KG. Kinetic and thermodynamic studies on fluoride biosorption by devdaru (polyalthia longifolia) leaf powder, Octa. Journal of Environmental Research. 2014;**2**:22-31

[141] Mohan SV, Ramanaiah SVV, Rajkumar B, Sarma PNN. Removal of fluoride from aqueous phase by biosorption onto algal biosorbent Spirogyra sp.-IO2: Sorption mechanism elucidation. Journal of Hazardous Materials. 2007;**141**:465-474

[142] Merugu R, Rao K, Garimella S, Medi NK, Prashanthi Y. Factors affecting the defluoridation of water using Fusarium oxysporum bioadsorbent. International Journal of Environmental Biology. 2013;**3**:12-14

[143] Ramanaiah SV, Venkata Mohan S, Sarma PN. Adsorptive removal of fluoride from aqueous phase using waste fungus (Pleurotus ostreatus 1804) biosorbent: Kinetics evaluation. Ecological Engineering. 2007;**31**:47-56

[144] Chhipa H, Acharya R, Bhatnagar M, Bhatnagar A. Determination of sorption potential of fermentation industry waste for fluoride removal. International Journal of Bioassays. 2013;**2**:568-574

[145] Mondal NK, Kundu M, Das K, Bhaumik R, Datta JK, Bengal W. Biosorption of fluoride from aqueous phase onto aspergillus and its calcium-impregnated biomass and evaluation of adsorption kinetics. Fluoride. 2013;**46**:239-245

[146] Mukkanti VB, Tembhurkar AR. Taguchi's experimental design for the optimization of the defluoridation process using a novel biosorbent developed from the clamshell waste. Journal of Dispersion Science and Technology. 2022;**0**:1-10

[147] Yan H, Wei M, Ma J, Li F, Evans DG, Duan X. Theoretical study on the structural properties and relative stability of M (II)-Al layered double hydroxides based on a cluster model. The Journal of Physical Chemistry. A. 2009;**113**:6133-6141

[148] Lu H, Li Q, Xiao H, Wang R, Xie D. Effect of non-thermal plasma modified NiAl layered double hydroxides on the removal of fluoride from aqueous solution. American Journal of Analytical Chemistry. 2014;**5**:547-558

[149] Sadik N, Mountadar M, Sabbar E. Defluoridation by calcined layered double hydroxides synthesized from seawater. Journal of Materials and Environmental Science. 2015;**6**:2239-2246

[150] Girma M, Zewge F, Chandravanshi BS. Fluoride removal from water using magnetic Iron oxide/aluminium hydroxide composite. SINET: Ethiopian Journal of Science. 2020;**43**:32-45

[151] Gitari WM, Izuagie AA, Gumbo JR. Synthesis, characterization and batch assessment of groundwater fluoride removal capacity of trimetal Mg/Ce/Mn oxide-modified diatomaceous earth. Arabian Journal of Chemistry. 2020;**13**:1-16

[152] Dessalegne M, Zewge F, Pfenninger N, Johnson CA, Diaz I. Layered double hydroxide and its calcined product for fluoride removal from groundwater of Ethiopian rift. Water, Air, & Soil Pollution. 2016;**227**:381-391

[153] Dang-bao T, Lam H, Dang T. Defluoridation of water by Ce-Ti hybrid oxide nanoparticles. IOP Conference Series: Earth and Environmental Science. 2021;**947**:1-6

[154] Cai P, Zheng H, Wang C, Ma H, Hu J, Pu Y, et al. Competitive adsorption characteristics of fluoride and phosphate on calcined Mg-Al-CO_3 layered double hydroxides. Journal of Hazardous Materials. 2012;**213-214**:100-108

[155] Kang D, Yu X, Tong S, Ge M, Zuo J, Cao C, et al. Performance and mechanism of Mg/Fe layered double hydroxides for fluoride and arsenate removal from aqueous solution. Chemical Engineering Journal. 2013;**228**:731-740

[156] Sun Z, Park J, Kim D. Synthesis and adsorption properties of Ca-Al layered double hydroxides for the removal of aqueous fluoride. Water, Air, & Soil Pollution. 2017;**228**:23-30

[157] Chen JP, Yu Y, Yu L, Chen JP. Adsorption of fluoride by Fe – Mg – La triple- metal composite: Adsorbent preparation, illustration of performance and study of ... preparation, illustration of performance and study of mechanisms. Chemical Engineering Journal. 2015;**262**:839-846

[158] Elhalil A, Qourzal S, Mahjoubi FZ, Elmoubarki R, Farnane M, Tounsadi H, et al. Defluoridation of groundwater

by calcined Mg/Al layered double hydroxide. Emerging Contaminants. 2016;**2**:42-48

[159] Das C, Dey U, Chakraborty D, Datta JK, Mondal NK. Fluoride toxicity effects in potato plant (solanum solanum tuberosum L.) grown in contaminated soils, Octa. Journal of Environmental Research. 2015;**3**:136-143

[160] Pitre D, Boullemant A, Fortin C. Uptake and sorption of aluminium and fluoride by four green algal species. Chemistry Central Journal. 2014;**8**:2-8

[161] Gandhi N, Sirisha D, Sekhar KBC. Phytoremediation of chromium and fluoride in industrial waste water by using aquatic plant Ipomoea aquatica. South pacific Journal of Pharma and Bio Science. 2013;**1**:1-4

[162] Parikh PS, Mazumder SK. Capacity of Azolla pinnata var. imbricata to absorb heavy metals and fluorides from the wastewater of oil and petroleum refining industry at Vadodara. International Journal of Allied Practice, Research and Review. 2015;**2**:37-43

[163] Braga AF, Borges AC, Vaz LRL, De Souza TD, Rosa AP. Phytoremediation of fluoride-contaminated water by Landoltia punctata. Engenharia Agrícola. 2021;**41**:171-180

Chapter 5

Sources of Human Overexposure to Fluoride, Its Toxicities, and Their Amelioration Using Natural Antioxidants

Thangapandiyan Shanmugam and Miltonprabu Selvaraj

Abstract

Fluoride (F) is released into the environment through a combination of natural and anthropogenic processes include the weathering from volcanoes, geothermal activity, and marine aerosols. Chronic fluoride exposure has been linked with amyriad of human diseases such as skeletal and dental fluorosis, diabetes, atherosclerosis, cardiovascular diseases, and hyperkeratosis. Since fluoride targets ubiquitous enzyme reactions, it affects nearly all organ systems in animals and humans. Apart from synthetic chemical chelators, studies have been carried out to explore natural antioxidants against F toxicity. Natural products contain substances that inhibit the theoxidation of substrate(s). Antioxidant molecules are thought to play a crucial role in counteracting free-radical-induced damage to macromolecules. In this book chapter literature survey of the different phytoremediation strategy is presented. The results show that natural antioxidants exhibit promising antidote against fluoride-induced toxicity in different mammal systems.

Keywords: amelioration, antioxidants, fluoride, natural products, overexposure, toxicity

1. Introduction

Trace elements such as Fluoride (F) are essential to animals and humans for normal health status. Fluorine is the ninth element on the periodic table. It has crustal abundance of 0.054%, which makes it the 24th most abundance element on the earth and most reactive member of the halogen family. It has an atomic weight of 18.9984. The physical and chemical properties of fluorine have been given in **Table 1**. Fluorine reacts with other elements to produce ionic compounds such as calcium fluoride (CaF), sodium fluoride (NaF), hydrogen fluoride (HF), aluminum fluoride (AlF), and many other compounds [1, 2]. In general, F-like elements causus only a source of local pollution [3]. The degree of toxicity, the scope of exploitation of the element, and its application and subsequent mobilization into the air, water, and soil are used to assess the environmental relevance of increased levels of these elements.

Physical properties	
Atomic number	9
Atomic mass	18.998403 g.mol^{-1}
Density	1.8×10^{-3} g.cm^{-3} at 20°C
Melting point	−219.6°C
Boiling point	−188°C
van der Waals radius	0.135 nm
Ionic radius	0.136 nm (−1); 0.007 (+7)
Isotopes	2
Chemical properties	
Electronegativity according to Pauling	4
Electronic shell	[He] $2s^2 2p^5$
Energy of the first ionization	1680.6 kJ.mol $^{-1}$
Energy of second ionization	3134 kJ.mol $^{-1}$
Energy of third ionization	6050 kJ mol^{-1}
Standard potential	−2.87 V
Discovered by	Moissan in 1886

Table 1.
The physical and chemical properties of fluorine.

2. Source human overeexposure to fluoride

2.1 Air

Even though F is extensively distributed in the environment, only a small portion of overall human fluoride exposure is through the air [4, 5]. This is because the F concentrations in the air are relatively low in non-industrial locations, but they rise steeply among industrial places where phosphate fertilizers are produced or fluoride-containing coal is burnt. As a result, human exposure to fluoride from ambient air has been estimated to be only about 1–4 mg/day [6]. This is insignificant compared with other sources of human fluoride exposure [7, 8]. Nonetheless, F can enter the air from sea spray, and therefore, the immediate atmosphere might be expected more enriched near or within the coastal areas [9]. No data were found on fluoride levels in ambient air or residential soil.

2.2 Water

F is generally prevalent in many water supplies and drinking water sources around the world because they leach into groundwater from F-containing rocks and soils [10]. Because drinking water is fluoridated artificially in certain regions, this is often the major contributor to daily F consumption by humans in those areas. It is found that children who drink 1 L of water per day may consume up to 1.2 mg of fluoride per day [9]. WHO (World Health Organization)'s maximum permissible limit of F in drinking water is 1.5 mg/L and highest desirable limit is 1.0 mg/L.Estimation of human lethal F shows a wide variation of values that range from 16 to 64 mg/kg in adults and from 3 to 16 mg/kg in children.

2.3 Toothpaste, mouthwash, and fluoride supplements

Over 80% of the toothpastes sold around the world are fluoridated, with fluoride concentrations ranging from 1 to 1.5 mg/g [11]. It is believed some fluorides are absorbed straight into the tooth enamel when a person brushes with fluoride toothpaste. Adults are estimated to ingest about 0.02–0.1 g of toothpaste per day; however, children may ingest 0.2–0.8 g per day [12]. Fluoride concentrations in mouthwashes range from 0.23 to 0.97 mg/gram [11]. Adults are likely to use and consume these more frequently than youngsters. On average, an adult person can swallow roughly 1.0 g of mouthwash every day, while a youngster would take about 0.5 g, according to the Office of Environmental Health Hazard Assessment (OEHHA).

2.4 Fluoride exposure from agricultural foodstuffs

Although raw foods contain some F, the greater portion of F exposure through the diet is as a result of F added to foods when they are cooked or processed with high F water. Nonetheless, among the foods that are highest in F include tea and ocean fish containing bones or bone meal. The consumption of tea in larger quantities can represent a potential F health risk because the tea plant (*Camellia sinensis*) is known to uptake high F levels from the soil and to accumulate them in the leaves, from where

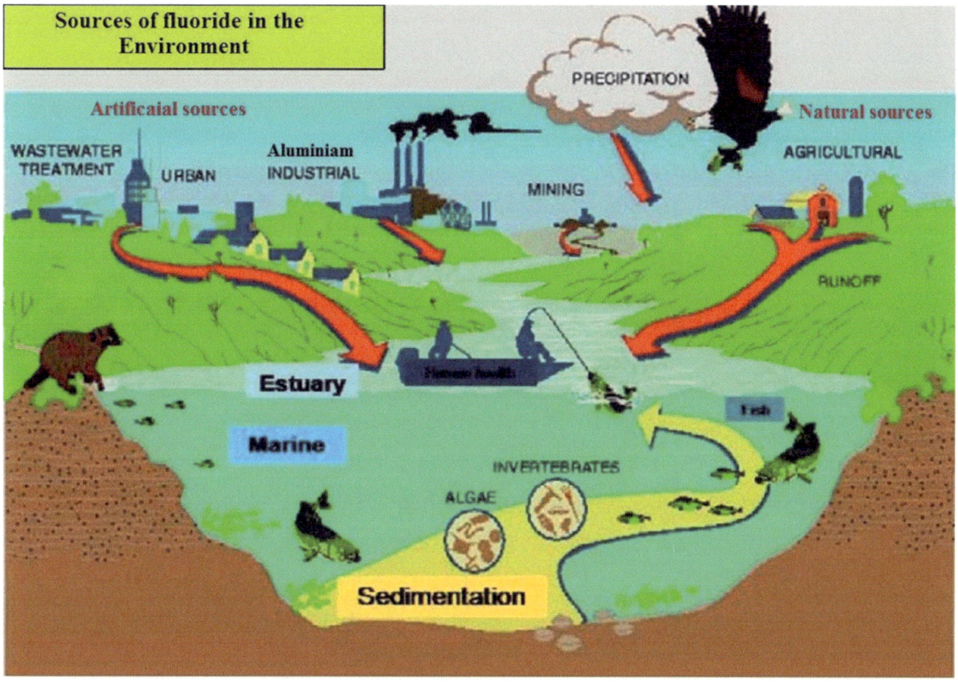

Figure 1.
Sources of fluoride and fluoride cycle in the environment.

it is easily released during the infusion of tea leaves during brewing. For communities with low F levels in foods, contribution in adults ranges from 0.3 to 1 mg/day.

2.5 Industrial source of fluoride pollution

Another major source of F pollution is the atmospheric or wastewater-based release from numerous industries that processor deal in steel, aluminum, copper, and nickel, phosphate ores, phosphatic fertilizers, glass, bricks, and other ceramics [4, 5]. The biggest industrial source of fluoride pollution into the environment is, however, the phosphate ore production and aluminum smelting. Fluoride dispersion, a pollution source, is also aided by the application of fluoride-containing insecticides and the combustion of coal and other fuel sources of geogenic origins. in the general environmental fluoride cycling between the surface waters, groundwater reserves, air, soils, and the biological components of the environment as a result of these activities is summarized in **Figure 1**.

3. Molecular mechanism of fluoride toxicity

F is redily absorbed by the stomach, lumen, and small intestine, and approximately 75–90% of ingested F is absorbed from the gastrointestinal tract. Fluoride

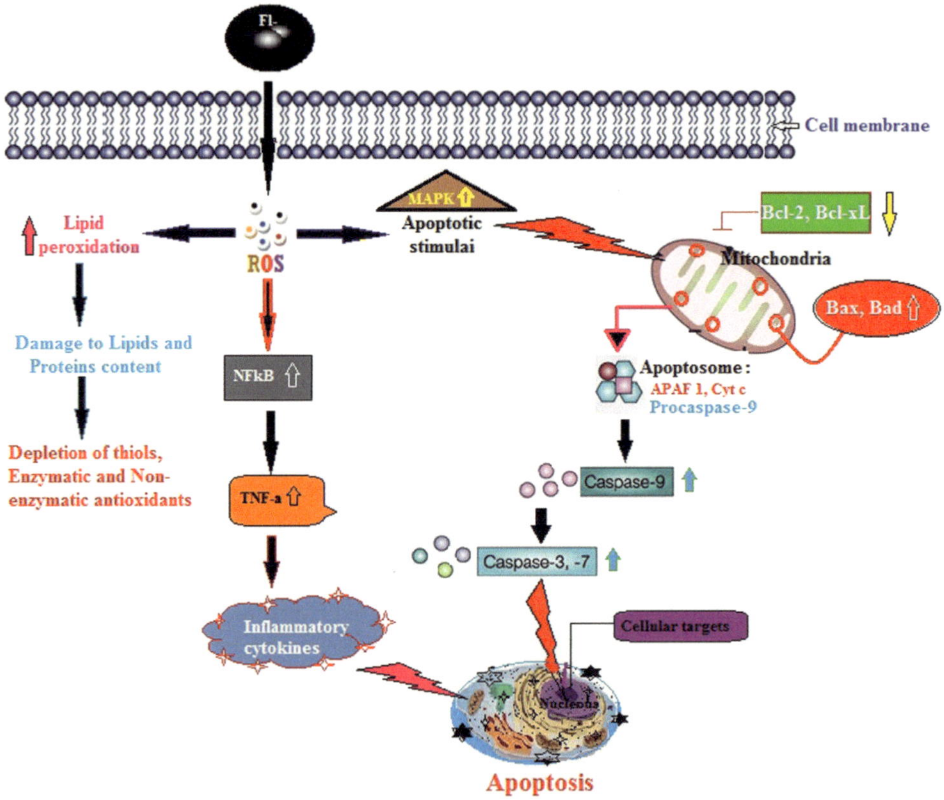

Figure 2.
Molecular mechanism of fluoride-induced toxicity.

transport through biological membranes occurs primarily through the non-ionic diffusion of hydrogen fluoride (HF). The small neutral molecule of HF penetrates cell membranes much faster than the dissociated fluoride ion, resulting in a more pronounced intracellular intake [13]. After ingestion, fluoride is rapidly and virtually absorbed into the blood stream. The ingested fluoride appears in the plasma within 30–60 min after uptake, and it is distributed from the plasma to all tissues and organs within 24 h. The element is taken up into all the tissues of the body, but it is retained and accumulated only in the teeth and other skeletal tissues resulting in dental and skeletal fluorosis when the elements water threshold levels of 1.5–20 mg/L are exceeded, respectively. In general, approximately 50% of absorbed F is retained by uptake in calcified tissues.

F is excreted primarily via urine. Urinary F clearance inceases with urine pH due to a decrease in the concentration of HF. The rate of F removal from plasma, which in healthy adults is approximately 75 mL/min, is approximately equal to the amount of the renal and calcified tissues clearances. For healthy young or middle-aged adults, only 50% of absorbed fluoride that is not assimilated into calcified tissues is excreted in the urine.

The intermediate interaction of fluoride with body systems between its absorption in the gut and its assimilation into skeletal tissue or renal clearance from the body results in a series of toxic effects to the body. These toxicities symptoms are generally referred to as non-skeletal fluorosis. It has been suggested that oxidative stress can be a possible mechanism through which fluoride induces damage to the various tissues. This F toxic mechanism can be summarized as in **Figure 2**.

4. Fluoride toxicity

4.1 Skeletal tissue toxicity

4.1.1 Dental fluorosis

Dental fluorosis is hypo-mineralization of teeth enamel that is characterized by greater surface and subsurface porosity than is found in normal enamel. It results from excess fluoride reaching the growing tooth during its developmental stages [14]. F has a greater affinity for developing enamel because tooth apatite crystals can bind and integrate fluoride ions into the crystal lattice [15]. Ameloblast epithelial cells are responsible for enamel development, and the life cycle of these cells has three stages, which comprise: secretory, transition, and maturation steps [16]. Tooth morphological studies [17] have shown that fluoride affects the secretory stage of the ameloblasts cells, which secrete the enamel proteins called enamelin, which mineralizes to form tooth enamel. When these stages are interfered with due to fluoride overexposure, the general mineralization of the enamel is compromised leading to dental fluorosis.

4.1.2 Skeletal fluorosis

Skeletal fluorosis is a painful crippling pathological condition, which can occur on long-term exposure to high dietary levels of fluoride exceeding 20 mg/L [18]. The mechanism of skeletal fluorosis suggests that fluoride ions are deposited in the bone by substituting hydroxyl groups in the carbonate apatite structure to produce fluorohydroxyapatite, thus altering the mineral structure of the bone according to the equation below:

$$Ca_{10}(PO_4)OH + F \rightarrow Ca_{10}(PO_4)OH_6F$$

Because, F altered the mineralization of bone strength and finally causes week bone or soft bone called skeletal fluorosis (**Figure 3**).

4.2 Non-skeletal/physiological toxicity

4.2.1 Gastrointestinal effect

The dominant form in which F exists in solution is highly pH-dependent. At the normal pH of drinking water (pH = 7), fluoride occurs primarily as the free ion, F^-. In the stomach, the ingested fluoride combines with hydrogen ions from HCl to form largely molecular HF, depending on the pH in the stomach (2.4% HF at pH 5; 96% HF at pH 2). The stomach is among the first target organs for the adverse effects of fluoride. Among the soft tissues of the body, the propensity of gastric mucosa exposed to the highest concentrations of the HF is immense. HF easily crosses the gastric epithelium and is the major form in which fluoride is absorbed from the stomach. Several functional and structural changes might be associated with ingestion of fluorides such as increased mucus secretion, followed by patchy or widespread loss of the mucus layer, hyperemia, edema, and hemorrhage [19].

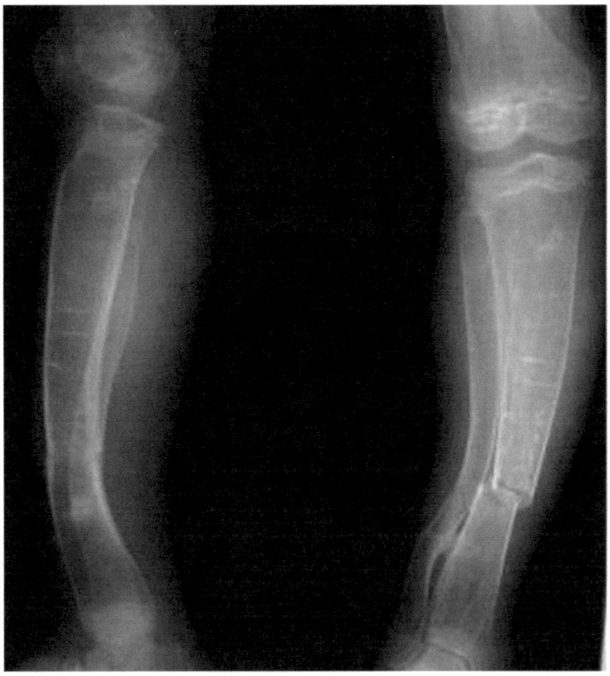

Figure 3.
Skeletal fluorosis shows brittle bone.

4.2.2 Hepatic effect

The liver is responsible for maintaining the body's metabolic homeostasis, and it has been considered as the key target organ for the toxic effects of fluoride [20]. Several mechanisms have been proposed to explain fluoride-induced hepatotoxicity. The important possible mechanism is the disturbance of prooxidant and antioxidant balances by the generation of reactive oxygen species (ROS). This decreases the activities of enzymatic antioxidants. Previous studies have shown fluoride induced abnormal function in the liver of rats, sheep, mice, etc. Especially, superoxide dismutase (SOD), catalase (CAT), and glutathione peroxidase (GPx) were decreased with increased lipid peroxidation product, which cause damage to hepatocytes [21].

4.2.3 Kidney effect

The kidney is the potential site of acute fluoride toxicity because kidney cells are exposed to relatively high F concentrations [22]. Fluoride concentrations in the kidney show an increase in the gradient. Therefore, the kidney is thought to be one of the target organs for the adverse effects of fluoride because of the bioconcentration, metabolism, and kinetics excretion. F-induced ROS increases the excessive generation of nitric oxide, oxygen-free radicals, decreased CAT, SOD, glutathione (GSH), and increased lipid peroxidation, which may lead to severe damages in the nephron structure and functions and also biomacromolecules, such as proteins and nucleic acids [22, 23].

4.2.4 Respiratory effect

Fluoride exposure has been associated with asthmatic symptoms among workers in the aluminum industry [24]. However, only recently there is mounting evidence that ROS plays an important part in the complex physiological processes such as cell signaling and apoptosis. One of the organs commonly affected by ROS generation is the lungs. It is obvious that having a large surface that is constantly in contact with atmospheric oxygen and pollutants, the lungs are a site of major ROS production. The most fundamental adverse effect of fluorides in the respiratory system is the inhibition of Kreb's cycle enzymes in the lung by subsequent production of ROS. Furthermore, several studies have shown that the interaction of the lung immune system and oxidative stress might be associated with the development of several other pulmonary or respiratory diseases [25].

4.2.5 Reproductive effect

One of the toxicants that have harmful effects on the reproductive system is fluoride. The metabolism and morphology of spermatozoa were altered in the fluoride-exposed rats due to enhanced ROS/RNS-mediated lipid peroxidation. Long et al. [26] reported that fluoride significantly declined the weights of testes and cauda epididymis in rats. Sialic acid is an important constituent of mucopolysaccharides and sialomucoproteins, which are essential for the maturation of spermatozoa in epididymis and maintenance of the structural integrity of their membranes. However, fluoride overexposure can significantly altered the sialic acid [27]. Sharma et al. [28] have reported that female rats exposed to 6 ppm concentrations of sodium fluoride for 15 and 30 days revealed that the reproductive organ weights of the ovary, uterus, and adrenal gland declined significantly due to the overproduction of reactive oxygen species with increased lipid peroxidation.

4.2.6 Cardiovascular effect

The heart is a muscular pumping organ, mainly involved in the purification and circulation of blood in the body. Heart failure results from a sudden reduction in coronary blood flow to a segment of the myocardium, which initiates severe cellular changes in the myocytes that, inevitably culminating in cell death and tissue necrosis. Sinha et al. [29] have shown that fluoride consumption causes ROS-mediated myocardium injuries and dysfunction. Also, Nabavi et al. [30] have reported that fluoride increases oxidative stress through abnormal biochemical parameters in the heart tissues of rats. Fluoride-induced oxidative stress plays an important role in the progression of a variety of cardiac disorders such as cardiac failure and ischemia [30]. Nicotinamide adenine dinucleotide phosphate-oxidase (NADPH oxidase) (Nox) is an important source of ROS in the vasculature and is activated by high levels of fluoride exposure. Fluoride can stimulate the Nox expression, and activity has implications in endothelial dysfunction and vascular disorders. It is possible that endothelial dysfunction in coronary heart disease could be related to the chronic inflammation that coexists with atherosclerosis [31].

4.2.7 Neurological effect

Fluoride is a powerful toxin of the central nervous system and adversely affects the brain functioning even at low doses [32]. It can induce neuron apoptosis and decreased cerebral functions, impaired memory and learning ability [33, 34]. F ions bind to antioxidants such as N-acetyl cysteine (NAC), glutathione (GSH), and other free-radical-defeating enzymes. This causes oxidative stress and eventually cell death [35]. The lack of a compensating antioxidant system combined with oxidative stress caused by increased free radicals plays a significant role in the onset of nerve cell membrane damage, particularly through enhanced lipid peroxidation. Several studies have reported the effects of fluoride in drinking water on cognitive capacities, and IQ reductions were observed at water-fluoride concentrations of about 1 mg/L and above [36].

5. Application of natural antioxidant against F toxicity in different organs

Studies have been carried out to explore natural antioxidants against toxic substances or elements [37, 38]. Antioxidant molecules are thought to play a crucial role in counteracting free-radical-induced damage to macromolecules [39]. There is a wide range of antioxidants that can counteract the condition of oxidative stress. It includes vitamins, phenolic compounds (flavonoids), and carotenoids. Minerals including selenium, zinc, manganese, magnesium, and copper also play a part in the body's hundreds of antioxidant functions. Aside from scavenging free radicals, natural antioxidants have been shown to influence the expression of several genes and signal regulatory pathways, potentially preventing cell death [40]. The natural antioxidants listed below were employed to alleviate F-induced oxidative stress-mediated damage in several organs in rats.

5.1 Hesperidin against F hepato toxicity

Recently, Caglayan et al. [41] reported that 600 ppm of fluoride administration induced the hepatotoxicity with severe damage in rats. However, hesperidins (HSP) 200 ppm administration to F group recovered completely from the F-induced toxicity in rats. The data showed the antioxidative potential of HSP on F-induced toxicity in liver tissue. Similarly, Küçükler et al. [42] also demonstrated the antioxidant property of HSP (100 mg/kg) in albino Wistar rats against F-induced hepatoxicity.

5.2 *Prunella vulgaris* against F nephrotoxicity

Natural antioxidants can help to conquer oxidative stress and free-radical-induced disorders. In an earlier study, epigallocatechingallate (EGCG) depicted ameliorative effects toward F nephrotoxicity in rats [43]. However, a recent study from Li et al., [44] observed that the oral administration of *P. vulgaris* (1.575 g crud drug) attenuated the fluoride-induced oxidative toxicity in rat kidneys. The hole plants were dried and powdered and extracted (Ethanol 80%, Distilled water 20%) the drug by using reflux extraction apparatus. This result indicates that natural antioxidants exhibit the attenuating efficacy over xenobiotics of the toxic element on different organs. After extraction of PV drug, it was passed through a series of HPLC (high-performance liquid chromatography), to identified the natural bioactive compounds followed by structural identification of natural bioactive compounds by using XRD.

5.3 Silymarin against F cardiotoxicity

Milk thistle extract has a centuries-old history of use in Indian folk medicine to treat a variety of illnesses including jaundice, gallstones, hemorrhage, bronchitis, or varicose veins. Its beneficial effects have been attributed to the antioxidant, anti-proliferative, and anti-inflammatory effects based on the regulation of specific signaling pathways. Nabavi et al. [45] reported that fluoride caused severe damage to the heart and brain tissue after administration. These effects are due to the alteration of enzymatic and non-enzymatic antioxidants with increased lipid peroxidation markers. This exploration shows that the Silymarin has the potential phytoremediation property against F-induced toxicity in rats.

5.4 Curcumin against F *in-vitro* toxicity

Curcumin longa is widely used as a food additive and was one of the bioactive natural products in phytoremediation therapy. In the traditional medicine of India, a cream of *Curcuma longa* called Ayurveda is used for the treatment of eye diseases, wounds, bites, burns, and various dermal diseases [46]. Fujiwara et al. [47] recently reported that the antioxidant efficacy of Curcumin in an in-vitro study against ameloblast LS8 cells treated with fluoride toxicity. In this study, Curcumin significantly attenuated the F toxicity on ameloblast LS8 cells due to the potent phytotherapeutic property of curcumin and its derivatives.

5.5 Quercetin against F neuro toxicity

Quercetin is one of the best-studied polyphenols found in onions, apples, berries, tea, and red wine. It prevents apoptosis, anticancer, anti-inflammatory, and cardiovascular protective ability via its potent antioxidant properties [48]. Chouhan et al. [49] reported that Silymarin and Quercetin synergistically abrogated fluoride-induced oxidative toxic effect in rat liver and kidney via activation of phase I antioxidant enzymes such as SOD, CAT, GPx. Similarly, Nabavi et al. [50] reported that Quercetin showed preventive effect of fluoride toxicity in the brain due to the high concentration of antioxidant capacity of Quercetin to abrogates the Fl-induced ROS in different organs.

6. Conclusions

Taken together, F elements causing many illness by producing oxidative stress-mediated toxicity in different organs. However, the natural antioxidant from plant sources has rich chelating ability than synthetic chelators for many uncurable diseases. The supplementation of natural antioxidants mentioned in this book chapter could mitigate all kinds of organ toxicity elicited by F via modulation of oxidative damage, apoptosis in both in-vitro and in-vivo studies on different animals. This exploration also shed the lights on natural antioxidant therapy and future research in this fields.

Acknowledgements

The authors would like to greatly acknowledge the Vice-Chancellor, Science Dean, and HOD of Zoology, Sri Ramasamy Memorial University, Sikkim, and Madras University, Chennai, India, for providing all the facilities and support for writing this book chapter.

Conflict of interest

The authors declared that there is "no conflict of interest."

Author details

Thangapandiyan Shanmugam[1*] and Miltonprabu Selvaraj[2]

1 School of Basic Sciences, Department of Zoology, SRM University, Sikkim, India

2 Department of Zoology, University of Madras, Chennai, India

*Address all correspondence to: s.thangapandiyanphd@gmail.com

References

[1] Shanthakumari D, Srinivasalu S, Subramanian S. Effects of fluoride intoxication on lipidperoxidation and antioxidant status in experimental rats. Toxicology. 2004;**204**:219-228

[2] Budisa N, Kubyshkin V, Schulze-Makuch D. Fluorine-rich planetary environments as possible habitats for life. Life. 2014;**4**(3):374-385

[3] Nabavi SF, Habtemariam S, Sureda A, Akbar HM, Daglia M, et al. In vivo protective effects of gallic acid isolated from peltiphyllum peltatum against sodium fluoride-induced oxidative stress in rat erythrocytes. Arhiv za Higijenu Rada i Toksikologiju. 2013;**64**:553-559

[4] WHO (World Health Organization). In: Bailey K, Chilton J, et al., editors. Fluoride in Drinking Water. Geneva, Switzerland: WHO Press; 2006

[5] Environmental Health Criteria (EHC) 227, Fluorides, World Health Organization, Geneva, Switzerland, 2002

[6] U.S. Public Health Service. Review of fluoride, benefits and risks, report of the ad hoc subcommittee on fluoride of the committee to coordinate environmental health and related programs. 1991

[7] Medvedeva V. Structure and function of the mucosa of the stomach and duodenum in aluminum smelter workers (abstract). GigeienaTruda i Professions L'nyeZabolevanija. 1983;(11):25-28

[8] Desai V, Bhavsar B, Mehta NR, Saxena DK, Kantharia SL. Symptomatology of workers in the fluoride industry and fluorospar processing plants, fluoride research 1985. Studies in Environmental Sciences, Elsevier Science Publishers BV, Amsterdam. 1986;**27**:193-199

[9] Fawell J, Bailey K, Chilton J, Dahi E, Fewtrell L, Magara Y. Fluoride in Drinking water. WHO Drinking-Water Quality Series. London, Seattle: IWA Publishing; 2006

[10] Agency for Toxic Substances and Disease Registry (ATSDR). Toxicological Profile for Fluorides, Hydrogen Fluoride, and Fluorine. Atlanta, US: US Department of Health and Human Services; 2003

[11] Whitford G. The metabolism and toxicity of fluoride. Monographs in Oral Science. 1989;**13**:1-160

[12] US EPA. National Primary Drinking Water Regulations; fluoride. U.S. Environmental Protection Agency. Federal Register. 1985;**50**(93):20164-20175

[13] Singer L, Ophaug R, Harland BF. Fluoride intakes of young male adults in the United States. The American Journal of Clinical Nutrition. 1980;**33**:328-332

[14] Fejerskov O, Manji F, Bælum V. The nature and mechanisms of dental fluorosis in man. Journal of Dental Research. 1990;**69**:692-700

[15] Robinson C, Kirkham J, Weatherell JA. Fluoride in teeth and bone. Fluoride in dentistry. In: Fejerskov O, Ekstrand J, Burt BA, editors. 2nd ed. Copenhagen: Munksgaard; 1996. pp. 69-87

[16] Matsuo S, Inai T, Kurisu K, et al. Influence of fluoride on secretory pathway of the secretory Ameloblast in rat incisor tooth germs exposed to

sodium fluoride. Archives of Toxicology. 1996;**70**:420-429

[17] Skobe Z. The secretory stage of Amelogenesis in rat mandibular incisor teeth observed by scanning electron microscopy. Calcified Tissue Research. 1976;**21**:83-103

[18] Shivarajashankara YM, Shivashankara AR, Gopalakrishna BP, Rao SH. Oxidative stress in children with endemic skeletal fluorosis. Fluoride. 2001;**34**:108-113

[19] Spark CJ, Sjostedt S, Eleborg L, et al. Tissue response of gastric mucosa after ingestion of fluoride. British Medical Journal. 1989;**298**:1686-1687

[20] Thangapandiyan S, Miltonprabu S. Molecular mechanism of fluoride induced oxidative stress and its possible reversal by chelation therapy. Research and Review: A Journal of Toxicology. 2013;**3**(2):1-11

[21] Whitford GM. The physiological and toxicological characteristics of fluoride. Journal of Dental Research. 1990;**69**(539-549):556-557

[22] Nabavi SF, Moghaddam AH, Eslami S, et al. Protective effects of curcumin against NaF induced toxicity in rat kidneys. Biological Trace Element Research. 2013;**145**:369-374

[23] Yu RA, Xia T, Wang AG, et al. Effects of selenium and zinc on renal oxidative stress and apoptosis induced by fluoride in rats. Biomedical and Environmental Sciences. 2006;**19**:439-444

[24] Soyseth V, Kongerud J. Prevalence of respiratory disorders among aluminiumpotroom workers in relation to exposure to fluoride. British Journal of Industrial Medicine. 1992;**49**:125-130

[25] MacNee W, Rahman I. Is oxidative stress central to the pathogenisis of chronic obstructive pulmonary disease? Trends in Molecular Medicine. 2001;**7**:55-62

[26] Long H, Jin Y, Lin M, Sun Y, Zhang L, Clinch C. Fluoride toxicity in the male reproductive system. Fluoride. 2009;**42**:260-276

[27] Rajalakshmi M, Prasad MRN. Contribution of the epididymis and vas deferens in maturation of spermatozoa. In: Talwar GP, editor. Recent Advances in Reproduction and Regulation of Fertility. Amsterdam: Elsevier; 1979. pp. 253-258

[28] Sharma JD, Solanki M, Solanki D. Sodium fluoride toxicity on reproductive organs of female albino rats. Asian Journal of Experimental Sciences. 2007;**21**(2):359-364

[29] Sinha M, Manna P, Sil PC. Terminaliaarjuna protects mouse hearts against sodium fluoride-induced oxidative stress. Journal of Medicinal Food. 2008;**11**(4):733-740

[30] Nabavi SF, Nabavi SM, Ebrahimzadeh MA, et al. The protective effect of curcumin against sodium fluoride-induced oxidative stress in rat heart. Archives of Biological Science. (Belgrade). 2011;**63**(3):563-569

[31] Wolf J, Dagher MC, Fuchs A, et al. *In vitro* activation of the NADPH oxidase by fluoride. Possible involvement of a factor activating GTP hydrolysis on Rac (Rac-GAP). European Journal of Biochemistry. 1996;**239**:369-375

[32] Lu XH, Li GS, Sun B. Study of the mechanism of neuron apoptosis in rats from the chronic fluorosis. Chinese Journal of Endemiology. 2000;**19**:96-98

[33] Sun ZR, Lie FZ, Wil N. Effects of high fluoride drinking water on the

cerebral function of mice. Chinese Journal of Endemiology. 2000;**19**:262-263

[34] Wang J, Ge Y, Ning H, et al. Effects of high fluoride and low iodine on biochemical indexes of the brain and learning-memory of offspring rats. Fluoride. 2004;**37**:201-208

[35] Anuradha C, Kanno S, Hirano S. Fluoride induces apoptosis by Caspase-3 activation in human leukemia HL cells. Archives of Toxicology. 2001;**74**:226-230

[36] Guan Z, Wang Y, Xiao K, et al. Influence of chronic fluorosis on membrane lipids in rat brain. NurtoxicolTera. l998;**20**:537-542

[37] Hassan HA, Abdel-Aziz AF. Evaluation of free radical-scavenging and anti-oxidant properties of blackberry against fluoride toxicity in rats. Food and Chemical Toxicology. 2010;**48**:1999-2004

[38] Ghosh J, Das J, Manna P. Cytoprotective effect of arjunolic acid in response to sodium fluoride mediated oxidative stress and cell death via necrotic pathway. Toxicology In Vitro. 2008;**22**:1918-1926

[39] Nair SB, Jhala DD, Chinoy NJ. Beneficial effects of certain antidotes in mitigating fluoride and arsenic induced hepatotoxicity in mice. Fluoride. 2004;**37**(2):60-70

[40] Zhao K, Zhao GM, Wu D, Soong Y, Birk AV, Schiller PW, et al. Cell-permeable peptide antioxidants targeted to inner mitochondrial membrane inhibit mitochondrial swelling, oxidative cell death, and reperfusion injury. Journal of Biological Chemistry. 2004;**279**(33):34682-34690

[41] Caglayan FMK, Darendelioğlu E, Ayna SKA. Hesperidin protects liver and kidney against sodium fluoride-induced toxicity through anti-apoptotic and anti-autophagic mechanisms. Life Sciences. 2021;**281**(15):119730

[42] Küçükler S, Çomaklı S, Özdemir S, Çağlayan C, Kandemir FM. Hesperidin protects against the chlorpyrifos-induced chronic hepato-renal toxicity in rats associated with oxidative stress, inflammation, apoptosis, autophagy, and up-regulation of PARP-1/VEGF. Environmental Toxicology. 2021;**36**(8):1600-1617

[43] Thangapandiyan S, Miltonprabu S. Epigallocatechingallate supplementation protects against renal injury induced by fluoride intoxication in rats: Role of Nrf2/HO-1 signaling. Toxicology Reports. 2014;**1**:12-30

[44] Li L, Lin LM, Deng J, Lin XL, Li YM, Xia BH. The therapeutic effects of *Prunella vulgaris* against fluoride-induced oxidative damage by using the metabolomics method. Environmental Toxicology. 2021;**36**(9):1802-1816. DOI: 10.1002/tox.23301

[45] Nabavi SM, Sureda A, Nabavi SF, et al. Neuroprotective effects of Silymarin on sodium fluoride-induced oxidative stress. Journal of Fluorine Chemistry. 2012;**142**:79-82

[46] Ammon H, Wahl MA. Pharmacology of curcuma longa. Planta Medica. 1991;**57**:1-7

[47] Fujiwara' Gary M. WhitfordJohn D. Bartlett Maiko Suzuki. Curcumin suppresses cell growth and attenuates fluoride-mediated Caspase-3 activation in ameloblast-like LS8 cells. Environmental Pollution. 15 Mar 2021;**273**:116495

[48] Boots AW, Haenen GR, Bast B. Health effects of quercetin: From antioxidant to nutraceutical. European Journal of Pharmacology. 2008;**585**:325-337

[49] Chohan S, Yadav A, Kushwah P, et al. Silymarin and quercetin abrogates fluoride induced oxidative stress and toxic effects in rats. Molecular & Cellular Toxicology. 2011;7:25-32

[50] Nabavi SF, Nabavi SM, Mirzaei M, Moghaddam AH. Protective effect of quercetin against sodium fluoride induced oxidative stress in rat's heart. Food & Function. 2012;**3**:437-441

Chapter 6

Ammonium Fluorides in Mineral Processing

Alexander Dyachenko

Abstract

The possibility of using ammonium fluoride as a new reagent for processing mineral raw materials is considered. Ammonium fluorides are the most convenient and technological fluorinating agents for the decomposition of the silicon component of ores. Advantages the use of ammonium fluoride (or hydrodifluoride) as a desiliconizing agent, the possibility of its complete regeneration. The processes of deep processing of silicon, zirconium, titanium, and beryllium minerals are considered. The excellence of using ammonium fluoride in the processing of mineral raw materials have been proven. The physicochemical laws of the processes are considered, technological schemes are proposed. The material will be useful in the further introduction of fluoride technologies at enterprises for the processing of quartz, zircon, ilmenite, and phenakite.

Keywords: mineral raw materials, ammonium fluoride, beryllium, zirconium, titanium dioxide, silicon dioxide

1. Introduction

Hydrometallurgy of non-ferrous and rare metals with a high energy intensity of the release of microcomponents from mineral raw materials and a significant negative situation on the biosphere. These factors lead to the need to create fundamentally new technologies in which the amount of waste is minimized. In an ideal chemical technology, any waste should become a commercial product, and the reagents should undergo complete regeneration and return to production.

Currently, one valuable component is usually extracted from complex ores, the rest go into slag. The amount of waste in modern chemical plants is boggles the imagination.

The basis of mineral and technogenic raw materials is usually silicon oxide and iron oxides. The opening of the silica component is chemically difficult, and the removal of a large amount of the cheap iron component can make the entire processing of the material unprofitable. Silicate minerals interfere with the hydrometallurgical interaction of the recovered commercial component with the reagent. Pyrometallurgical technologies are energy consuming.

One of the promising technologies is the fluoroammonium technology for processing mineral raw materials.

This article is devoted to the technology of complex processing of natural mineral and technogenic raw materials using ammonium fluorides.

The difference in the properties of ammonium fluorometallates is the physicochemical basis of the decomposition of mineral raw materials using ammonium fluorides. Metal fluorides notably differ in their in boiling point. Some are volatile and evaporate or sublimate when heated, separating from the main mass. Other fluorides are soluble and can be leached out of the fluorinated mass. Some undergo pyrohydrolysis or have different precipitation pH. After fluorination in a molten ammonium fluoride, a mixture of fluorides and ammonium fluorometallates is obtained.

By varying the differences in the physicochemical properties of ammonium fluorides and fluorometallates, it is possible to select modes for the complete separation of the mineral mixture into individual components.

Ammonium fluoride NH_4F under normal conditions is a non-aggressive, solid, and crystalline substance. Molten ammonium fluoride is an energetic fluorinating agent. Instead of ammonium fluoride, it is possible to use ammonium bifluoride NH_4HF_2. The melting point of NH_4F is 126°C, the boiling point of NH_4HF_2 is 239°C. NH_4HF_2 vapors are mainly composed of HF and NH_3. NH_4HF_2 is highly soluble in water, anhydrous HF, and hydrofluoric acid.

The interaction of most oxides with elemental fluorine, hydrogen fluoride, and hydrofluoric acid has been studied in sufficient detail [1]. The fluorination of mineral raw materials with ammonium fluorides requires further study. Silicon oxide is removed from the system in the form of volatile ammonium hexafluorosilicate at temperatures above 320°C.

Fluorination of metal oxides with fluoride and ammonium hydrodifluoride has been studied to a lesser extent. And the reactions of some fairly common oxides (Sn, Cu, Mn, and Ca), in order to create technological processes, have practically not been studied. The desiliconization cycle makes it possible to project similar processes of fluorination and regeneration of ammonium fluoride onto other oxides.

Regeneration is carried out due to the fact that oxides react well with molten ammonium fluorides at elevated temperatures, but do not react with ammonium fluoride solution in an alkaline medium.

2. Methodology

Thermal analysis was performed using thermogravimetric analysis (DTA)—is the starting characterization test for any thermal analysis. This characterization test gives an understanding of the thermal stability of a sample by giving a weight loss/gain signal as the sample is heated at a known rate in time and exposure to a given atmosphere. Hydrofluorination of oxides in a molten ammonium fluoride was carried out using a TGA/DSC/DTA analyzer SDT Q600 with software processing of data from TA Universal V4.2E instruments. Sample weight: up to 200 mg. Thermocouples: Pt/Pt-Rh. Crucibles: platinum, volume 110 μl. The temperatures of the onset of the reaction, the formation, and decomposition of complex fluoroammonium salts were investigated by the DTA-methods [2].

Thermal analysis of the interaction of oxides with ammonium hydrodifluoride made it possible to determine the temperature ranges in which complex fluoroammonium complexes are formed and their thermal destruction to individual oxides occurs. SiO_2, Al_2O_3, Fe_2O_3, and TiO_2 form complex fluoroammonium compounds with ammonium fluorides—ammonium hexafluorometallates. NiO form ammonium tetrafluorometalate. CaO, CuO, and KOH—fluorinated to simple fluorides. Thermogravimetric analysis revealed the regularities of oxide fluorination, which are necessary for solving technological problems. The kinetic experiment made it possible to determine the activation energy and the reaction rate constant.

It is calculated that the thermodynamically optimal temperature for hydrofluorination of multicomponent silicate mixtures is 500 ± 20 K. At a lower temperature, chemical reactions slow down. Increasing the temperature is not advisable, because decomposition of ammonium bifluoride into gaseous ammonia and hydrogen fluoride.

Based on DTA, the decomposition temperatures of fluoride compounds Al, Fe, Ni, Mn, Ca, and Cu, formed as a result of hydrofluorination in the melt of ammonium bifluoride, were found. The formation temperature of aluminum fluoride is 355°C, calcium fluoride 240°C, manganese fluoride 215°C, iron fluoride 365°C, nickel fluoride 295°C, and copper fluoride 260°C [3–5].

The kinetics of chemical reactions was investigated using the method of weighing the reacting mixture in the course of a chemical reaction. Weight loss occurs due to the formation of gaseous ammonia and water. The processing of experimental data was carried out according to the well-known methods of formal heterogeneous kinetics [6].

The possibility of separating multicomponent oxide silicate mixtures into individual oxides using only ammonium fluoride as an opening reagent has been experimentally proved.

Below are some specific examples of the application of ammonium fluoride technologies for the processing of mineral raw materials.

3. Results and discussion

3.1 Silicon dioxide processing

High purity silicon dioxide (SiO_2 content over 99.999%) is used in the production of optical glasses, optical fiber for Internet networks, silicon for solar energy, and electronics. The market for silicon dioxide is constantly growing and the demand for high-purity grades of silicon dioxide is especially high.

The raw material for the production of silicon dioxide is SiO_2-mineral and concentrate or quartz sand. The existing technologies for the production of synthetic silicon oxide are energy-consuming, multi-stage, and do not meet modern environmental requirements. We consider that the most promising direction is the fluoride technology for processing quartz raw materials using ammonium fluoride.

The advantages of NH_4F and NH_4HF_2 are the vigorous (energetic) interaction of the melt with silicon oxide, thus forming solid $(NH_4)_2SiF_6$ [7]. When heated, $(NH_4)_2SiF_6$ sublimes without decomposition, and when cooled, it desublimes. Multiple sublimation-desublimation is used for deep purification of quartz concentrate from impurities [8].

Ammonium fluoride reacts with the original mineral quartz sand according to the reaction:

$$SiO_2 + 6NH_4F \rightarrow (NH_4)_2SiF_6 + 2H_2O + 4NH_3 \quad (1)$$

Ammonium bifluoride reacts with silicon oxide according to the reaction:

$$SiO_2 + 3NH_4HF_2 \rightarrow (NH_4)_2SiF_6 + 2H_2O + NH_3 \quad (2)$$

The $(NH_4)_2SiF_6$ formed as a result of the reaction turns into a gaseous state when heated. Gaseous $(NH_4)_2SiF_6$ is condensed and treated with ammonia water with associated regeneration of the fluorinating agent. This process is described by the reaction:

$$(NH_4)_2SiF_6 + 4NH_4OH \rightarrow SiO_2 + 6NH_4F + 2H_2O \quad (3)$$

Next, the precipitate of hydrated silicon oxide is separated by filtration from the ammonium fluoride solution. The separated solution of ammonium fluoride is evaporated and crystallized in the form of technical ammonium fluoride of the composition 25% NH_4F and 75% NH_4HF_2. As a result of drying and calcining the precipitate, silicon oxide is obtained in a finely dispersed form.

The process is clearly displayed so-called fluoroammonium cycle (**Figure 1**).

Quartz sand with a known impurity content was used as a raw material (**Table 1**).

When the raw material interacts with NH_4HF_2, the compounds of impurity elements form the following fluorides:

$$Al_2O_3 + 6NH_4HF_2 \rightarrow 2(NH_4)_3AlF_6 + 3H_2O \quad (4)$$

$$Fe_2O_3 + 6NH_4HF_2 \rightarrow 2(NH_4)_3FeF_6 + 3H_2O \quad (5)$$

$$TiO_2 + 3NH_4HF_2 \rightarrow (NH_4)_2TiF_6 + 2H_2O + NH_3 \quad (6)$$

$$CaO + NH_4HF_2 \rightarrow CaF_2 + H_2O + NH_3 \quad (7)$$

$$MgO + NH_4HF_2 \rightarrow MgF_2 + H_2O + NH_3 \quad (8)$$

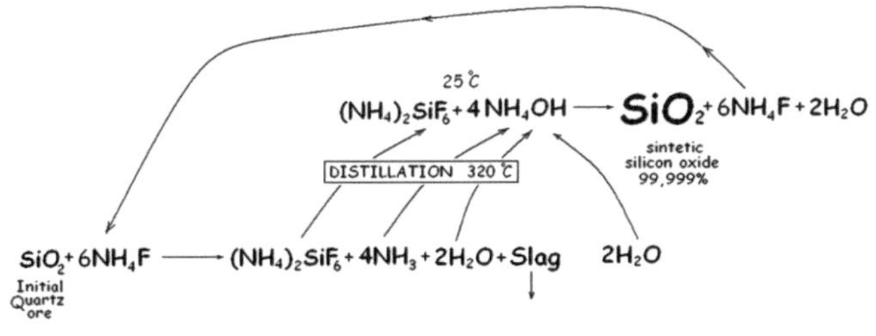

Figure 1.
Schematic diagram of fluoroammonium purification of silicon dioxide.

Elements,%							
SiO_2	AI	Ti	Na	Mg	K	Ca	Fe
97.5	0.28	0.11	0.06	0.06	0.1	0.84	0.96

Table 1.
Composition of raw materials (quartz sand).

Side reactions of the formation of nonstoichiometric silicon fluorides:

$$5(NH_4)_2SiF_6 + SiO_2 \rightarrow 6NH_4SiF_5 + 4NH_3 + 2H_2O \quad (9)$$

$$SiO_2 + 4NH_4HF_2 \rightarrow (NH_4)_2SiF_6 + 2H_2O + NH_3 + HF \quad (10)$$

$$NH_4SiF_5 + H_2O \rightarrow NH_4SiOF_3 + 2HF \quad (11)$$

It is known that reaction (9) can occur only at temperatures above 180°C. Therefore, upon fluorination, in our case, reaction (9) does not proceed. This reaction takes place in the next apparatus with sublimation purification of $(NH_4)_2SiF_6$ from impurities. Upon dissolution and subsequent precipitation of $(NH_4)_2SiF_6$ with a 25% ammonia solution, an undesirable hard-to-filter silica gel is formed due to the presence of NH_4SiOF_3.

$$NH_4SiOF_3 + H_2O \rightarrow SiO_2(gel) + NH_4F + 2HF \quad (12)$$

In the sublimator $(NH_4)_2SiF_6$ evaporates and decomposes according to the reaction:

$$(NH_4)_2SiF_6 \rightarrow 2NH_3 + 2HF + SiF_4 \quad (13)$$

In the desublimator NH_3, HF, and SiF_4 are cooled to form $(NH_4)_2SiF_6$

$$2NH_3 + 2HF + SiF_4 \rightarrow (NH_4)_2SiF_6 \quad (14)$$

Experimentally, we noticed that the temperature of the desublimation process strongly affects the quality of the resulting desublimate, in particular, the ratio of the amount of ammonium fluoride and $(NH_4)_2SiF_6$, as well as the amount of impurities in condensed $(NH_4)_2SiF_6$. A series of experiments was carried out to determine the effect of temperature.

The freeze-drying process consists of two stages:

1. from 110 to 280°C—side reactions. Capturing NH_4HF_2, NH_4F, NH_4SiF_5, and NH_4SiOF_3. Removal of excess NH_3.

2. from 280 to 380°C—sublimation and capture of $(NH_4)_2SiF_6$.

To determine the thermal properties of the compounds formed as a result of hydrofluorination in the ammonium fluoride melt, and the temperatures of their decomposition, DTA were carried out (**Figure 2**).

The initial temperature of weight loss is equal to 100°C, the change in weight stops at a temperature of 252°C, 16% of the total weight of the sample remains not flown away. This residue is fluoride of silica sand impurities. The DTA graph shows two exothermic peaks with maximums at 152 and 243°C.

The second peak characterizes the sublimation of $(NH_4)_2SiF_6$ and NH_4HF_2. Using TA instruments Universal V4.2E, the enthalpies of these processes were calculated in the first case ΔH = 214 J/g, in the second case ΔH = 1547 J/g. Heat of the sublimation process: Q = −1761 J/g.

The feedstock (quartz sand) and the reagent (ammonium fluoride) are mixed in the mixer screw and fed to the rotary drum kiln. In the furnace, a chemical reaction of interaction between quartz and ammonium fluoride takes place. The formation of

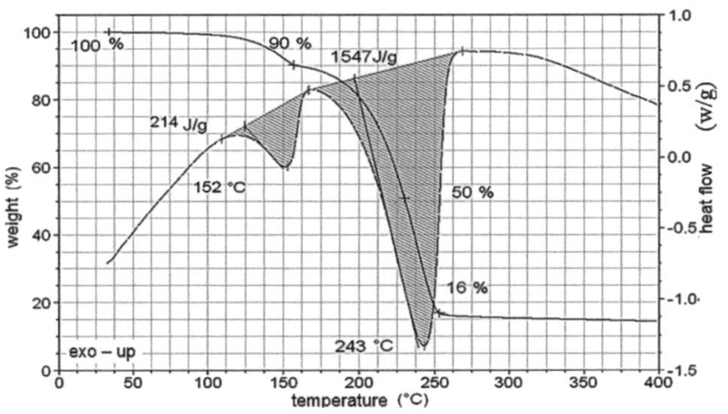

Figure 2.
Thermogravimetric and differential thermal analyzes of the decomposition of a fluorinated product. Heating rate 10°C/min.

solid primary $(NH_4)_2SiF_6$, gaseous water and ammonia is observed at a temperature of 200–220°C. The formed $(NH_4)_2SiF_6$ is heavily contaminated. It contains unreacted quartz and impurity fluoridation products. Impurities of Al, Fe, Ca, and many other substances are always contained in the original quartz sand. The gaseous phase containing ammonia and water vapor enters the absorption stage to produce ammonia water.

The solid phase [primary $(NH_4)_2SiF_6$] goes to the stage of sublimation purification in the next furnace. In a sublimation oven at a temperature of 320–350°C, gaseous $(NH_4)_2SiF_6$ evaporates. Impurities remain solid. Thus, the product is purified from impurities. The design of the sublimation oven is important. It is necessary to ensure high productivity of the process, but to prevent the ingress of impurities into the gas phase. Impurities can enter the gas phase due to the high velocity of the gas flow or due to intensive mixing of the reaction mass and the formation of dust. We suggest using a fixed bed furnace to prevent dust and impurities from entering the vaporized $(NH_4)_2SiF_6$. Gaseous $(NH_4)_2SiF_6$ from the sublimation furnace enters the condenser, where the gas is cooled and solid $(NH_4)_2SiF_6$ condenses. Sublimation and desublimation operations allow for high purity $(NH_4)_2SiF_6$. The impurity content can be reduced to 1 ppm. High purity $(NH_4)_2SiF_6$ dissolves in water.

Ammonia water is added to the solution and silicon oxide is precipitated. Regeneration of ammonium fluoride occurs as a result of the reaction of interaction of $(NH_4)_2SiF_6$ with ammonia water. The obtained silica precipitate is filtered to separate the ammonium fluoride solution. Silicon oxide is calcined in an oven to remove moisture. The ammonium fluoride solution is evaporated and crystallized. The regenerated ammonium fluoride again enters the stage of decomposition of a new portion of quartz sand. The hardware diagram of the experimental section consists of a number of standard and specially designed chemical devices (**Figure 3**).

As a result of studying the process of obtaining high-purity silicon oxide, the optimal conditions for the process of sublimation purification of $(NH_4)_2SiF_6$ were determined. In the temperature range from 110 to 280°C NH_4HF_2, NH_4F, NH_4SiF_5, and NH_4SiOF_3 are evaporated and excess NH_3 is removed. At temperatures from 280 to 380°C—sublimation and capture of $(NH_4)_2SiF_6$. It has been determined that at a desublimation temperature of 110–120°C it is possible to obtain the purest product with the highest content of $(NH_4)_2SiF_6$. The studies carried out made it possible to launch a pilot production of high-purity synthetic silicon oxide with a basic substance SiO_2 content of 99.999%.

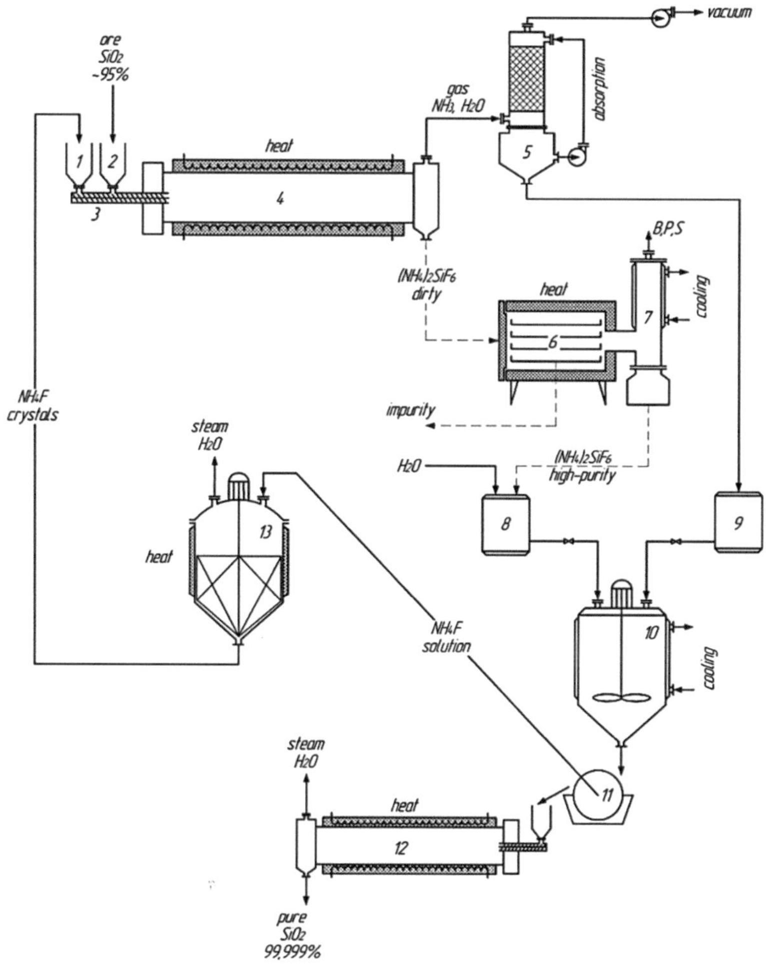

Figure 3.
Hardware diagram of fluoroammonium production of silicon dioxide. (1) Bunker for loading ammonium fluoride, (2) bunker for loading raw materials, (3) mixer screw, (4) drum rotary kiln, (5) ammonia absorber, (6) sublimation furnace, (7) condenser, (8) dissolution tank $(NH_4)_2SiF_6$, (9) tank for storing ammonia water, (10) reactor for precipitation of SiO_2, (11) vacuum filter, and (12) oven for drying SiO_2.

3.2 Zircon processing

A method for the autoclave decomposition of zircon with ammonium fluorides with the aim of producing zirconium oxide has been proposed and investigated.

It is known that zircon ($ZrSiO_4$) is one of the most chemically strong compounds. The molten ammonium fluoride at atmospheric pressure weakly interacts with zircon according to the reactions [9, 10]:

$$ZrSiO_4 + 13NH_4F \rightarrow (NH_4)_3ZrF_7 + (NH_4)_2SiF_6 + 8NH_3 + 4H_2O \quad (15)$$

$$2ZrSiO_4 + 13NH_4F \cdot HF \rightarrow 2(NH_4)_3ZrF_7 + 2(NH_4)_2SiF_6 + 3NH_3 + 8H_2O \quad (16)$$

Our studies have shown that fluorides react well with zircon at elevated pressures, that is, in an autoclave.

Having carried out a series of experiments on the decomposition of zircon with ammonium fluorides under various conditions, it was possible to find the optimal parameters that allow for a complete opening of the mineral and the conversion of zirconium into a soluble compound. It has been proven that the decomposition of zircon is faster when using ammonium bifluoride.

From the experimental data presented, it is possible to propose the optimal parameters of the process—the degree of response of more than 95% is achieved at a temperature of 300°C in 4 hours and at a temperature of 400°C in 1 hour.

The regeneration of ammonium fluoride provides a high economic attractiveness of the process and environmental safety. The resulting ammonium hexafluorosilicate sublimes at temperatures above 320°C and is removed from the mixture. As the temperature rises, $(NH_4)_2ZrF_6$ decomposes to zirconium tetrafluoride with the release of ammonia and hydrogen fluoride. The scheme of regeneration of ammonium fluoride and ammonium bifluoride is shown in **Figure 4**.

According to the proposed method, zircon, crushed to a particle size of 0.1 mm, is alloyed with ammonium bifluoride under isochoric conditions at a temperature of 300°C for 4 hours, while a pressure of up to 40 atm develops in the autoclave.

After decomposition, by the method of sublimation separation, zirconium tetrafluoride is isolated and purified. From the obtained ZrF_4 with the help of ammonia, zirconyl hydroxide—$ZrO(OH)_2$ is isolated.

3.3 Titanium dioxide processing

Titanium dioxide is one of the twenty main products of the chemical industry and is used as a white pigment in paints and varnishes. The ammonium fluoride method makes it possible to isolate titanium tetrafluoride from ilmenite $FeTiO_3$ in one stage and convert it into the form of titanium dioxide [11]. The interaction of ilmenite with ammonium fluoride proceeds with the formation of ammonium hexafluorotitanate and ammonium pentafluoroferrate according to reaction (17).

$$FeTiO_3 + 11NH_4F \rightarrow (NH_4)_2TiF_6 + (NH_4)_3FeF_5 + 6NH_3 + 3H_2O \quad (17)$$

This reaction begins at the melting temperature of ammonium fluoride—125°C, at a temperature of 280°C $(NH_4)_2TiF_6$ decomposes to TiF_4. At the same time, $(NH_4)_3FeF_5$ undergoes oxidation by atmospheric oxygen with simultaneous pyrohydrolysis according to reaction (18).

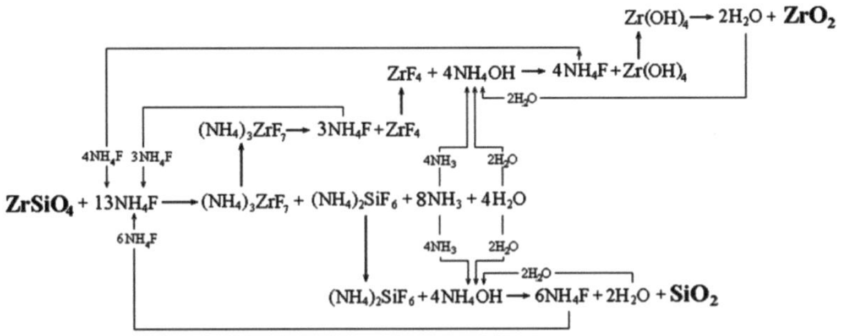

Figure 4.
Scheme of the decomposition of zircon into zirconium and silicon oxides.

$$4(NH_4)_2TiF_6 + 4(NH_4)_3FeF_5 + 4H_2O + O_2 \rightarrow 4TiF_4 + 2Fe_2O_3 + 20NH_4F + 8HF \quad (18)$$

Volatile titanium tetrafluoride, ammonia, water, ammonium fluoride at temperatures above 280°C are separated from iron(III) oxide. The kinetics of the fluorination of ilmenite with ammonium fluorides was studied experimentally.

In the temperature range 125–150°C, the activation energy of the process is 69 kJ/mol. The process takes place in the kinetic region of the reaction. The limiting stage of the process is the interaction of the reagents. The dependence of the degree of response on temperature and time is written by the Crank-Ginstling-Brounstein's equation:

$$\alpha = 1 - \left[1 - \sqrt{3,18 \cdot 10^4 \cdot e^{-\frac{68988}{R \cdot T}} \cdot \tau}\right]^3 \quad (19)$$

In the range 175–250°C, the activation energy of the process is 11 kJ/mol. The process takes place in the diffusion reaction region and is limited by the diffusion of the reaction products. In this interval, the degree of reaction can be determined by the Crank-Ginstling-Brounstein's equation:

$$\alpha = 1 - \left[1 - \sqrt{3, \cdot 10^{-3} \cdot e^{-\frac{11078}{R \cdot T}} \cdot \tau}\right]^3 \quad (20)$$

Ammonium hexafluorotitanate under the influence of temperature decomposes into gaseous titanium tetrafluoride, ammonia, hydrogen fluoride, and water. The gases are captured and interact when cooled according to reaction (21).

$$TiF_4 + 4NH_4OH \rightarrow Ti(OH)_4 + 4NH_4F \quad (21)$$

Ammonium fluoride is regenerated and titanium hydroxide is precipitated. **Figure 5** is a diagram showing the chemistry of the process and clearly depicting the return of ammonium fluoride to the cycle.

The only consumable reagent in the fluoroammonium processing of ilmenite is air oxygen, which is necessary for the oxidation of iron to the trivalent state. According to stoichiometry, 5.26 kg of oxygen is required for the oxidation of iron in 100 kg of ilmenite, which corresponds to 37 m^3 of air.

A schematic process flow diagram for producing titanium dioxide and iron oxide from ilmenite is shown in **Figure 6**.

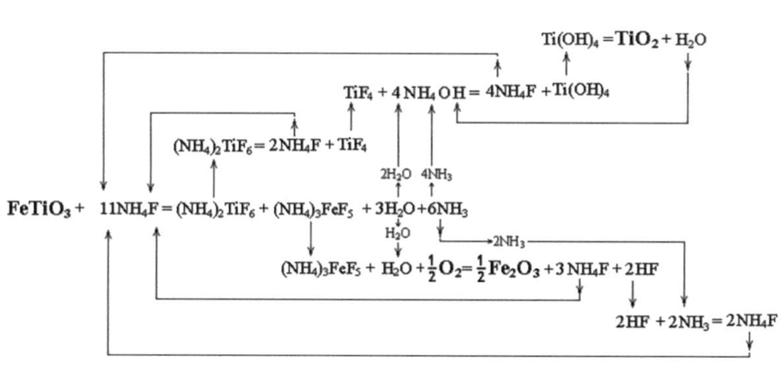

Figure 5.
Scheme of decomposition of ilmenite to titanium dioxide and iron oxide.

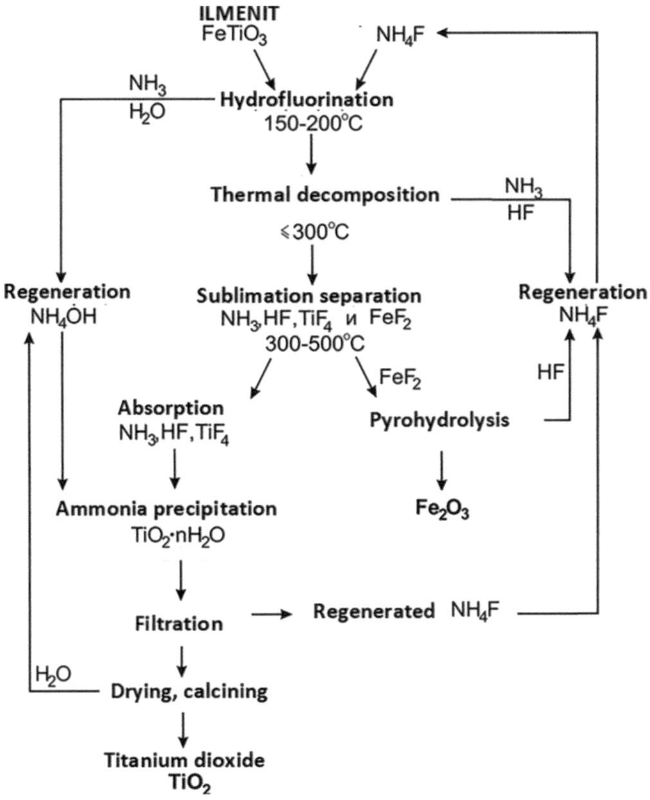

Figure 6.
Schematic diagram of fluoroammonium processing of ilmenite.

Ilmenite concentrate undergoes hydrofluorination in a molten ammonium fluoride at 150–200°C. In this case, fluoroammonium complexes of titanium and iron are formed, which decompose into iron difluoride and titanium tetrafluoride at temperatures above 300°C. After sublimation of titanium tetrafluoride, iron difluoride undergoes oxidative pyrohydrolysis with the formation of iron (III) oxide. Titanium tetrafluoride, separated from iron fluorides and impurities, is captured and precipitated by ammonia water to form hydrated titanium dioxide and ammonium fluoride solution.

After filtration, washing, drying and calcination of the resulting precipitate, titanium dioxide is obtained.

Based on the research carried out, a technological sequence of operations was developed. The modes of obtaining pigment TiO_2 and Fe_2O_3 from $FeTiO_3$ shown in **Table 2** were determined on a pilot batch of ilmenite concentrate.

A technique has been developed for processing ilmenite concentrate with ammonium fluoride to pigment titanium dioxide and iron (III) oxide. The optimal technological modes of fluoroammonium processing of ilmenite to TiO_2 (with the structure of rutile and anatase) and iron(III) oxide have been determined.

3.4 Beryllium processing

This chapter also describes the possibility of applying the fluorine-ammonium technology to the processing of beryllium ores—bertrandite $Be_4[Si_2O_7](OH)_2$ and phenakite Be_2SiO_4 and beryl BeO [12–14].

Stage (process)	Temperature, °C	Time, h	Product yield, %
Decomposition of ilmenite	150–200	1.5	95–99
Sublimation department TiF_4	400–500	1	95–98
TiF_4 capture	25	—	98.5
Sedimentation	25–35	0.5	99
Filtration	—	—	—
Anatase not less than 95%	500	2	—
Rutile not less than 95%	800	2	—
Pyrohydrolysis to Fe_2O_3	500	2	95
NH_4F regeneration (solution evaporation)	100–110		95

Table 2.
Modes of obtaining pigment titanium dioxide from ilmenite.

The thermodynamic probability of the reactions of interaction of the components of the beryllium concentrate with ammonium fluoride was calculated (**Table 3**).

The reaction of fluorination of phenakite with ammonium fluoride at temperatures above 500 K proceeds in the forward direction with the formation of ammonium tetrafluoroberyllate and ammonium hexafluorosilicate. The reaction of phenakite fluorination with ammonium hydroditoride begins already at a temperature of 127°C, with an increase in temperature, the reaction proceeds more fully with the formation of products.

The reaction begins to proceed at room temperature with the formation of the beryllium fluoroammonium complex $(NH_4)_2BeF_4{\cdot}nNH_4F$ and with the release of gaseous ammonia and water. With further heating, the process of decomposition of the fluoroammonium complex of beryllium proceeds. At 200°C $(NH_4)_2BeF_4{\cdot}NH_4F$ is formed, which, when heated to 240°C, decomposed to $(NH_4)_2BeF$. When the temperature rises to 280°C, $(NH_4)_2BeF_4$ decomposes to NH_4BeF_3, which, in turn, decomposes to BeF_4 at 385°C.

Based on the analysis of thermogravimetric studies, it is possible to propose the following chain of chemical transformations occurring during the interaction of beryllium oxide with fluoride and ammonium hydrodifluoride (20):

$$\begin{aligned} &BeO + NH_4F \cdot HF \underset{\rightarrow}{200°C} (NH_4)_2BeF_4 \\ &\cdot NH_4F \underset{\rightarrow}{240°C} (NH_4)_2BeF_4 \underset{\rightarrow}{280°C} NH_4BeF_3 \underset{\rightarrow}{385°C} BeF_2 \end{aligned} \quad (22)$$

The decomposition kinetics of beryllium oxide is shown in **Figure 7**.

T, °C	25	127	227	327	427	527
$Be_2SiO_4 + 14NH_4F \rightarrow 2(NH_4)_2BeF_4 + (NH_4)_2SiF_6 + 8NH_3 + 4H_2O$						
ΔG, кДж/моль	20	−181	−384	−588	−793	−999
$Be_2SiO_4 + 7NH_4HF_2 \rightarrow 2(NH_4)_2BeF_4 + (NH_4)_2SiF_6 + NH_3 + 4H_2O$						
ΔG, kJ/mol	−224	−297	−353	−394	−424	−443

Table 3.
Thermodynamics of the process of phenakite fluorination with ammonium fluorides.

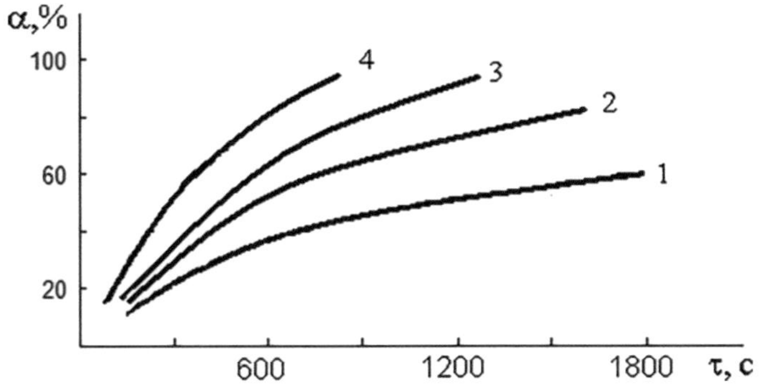

Figure 7.
Kinetics of interaction of beryllium oxide with ammonium fluoride. (1) 140°C, (2) 160°C, (3) 180°C, and (4) 200°C.

The kinetic equation for describing the rate of the process is found experimentally:

$$\alpha = 1 - \left(1 - 1,8 \cdot e^{-\frac{31000}{8,31 \cdot T}} \cdot \tau\right)^3 \tag{23}$$

The activation energy of the process was 31 kJ/mol, which indicates the occurrence of the reaction in the transition region between diffusion and kinetic. Kinetic studies have shown that in 20 minutes at a temperature of 200°C, the decomposition of beryllium oxide occurs by more than 95%.

Below is a diagram of a closed fluoroammonium cycle of decomposition of phenakite into silicon oxide and beryllium oxide with the regeneration of ammonium fluoride. The diagram in **Figure 8** clearly illustrates the closure of flows and the equality of material balance.

Thermodynamic and thermal analyzes of the considered system, which proved the theoretical and laboratory feasibility of the process, made it possible to proceed to the development of the process flow diagram (**Figure 9**).

The original fluorite-phenakite concentrate, containing 30% phenakite, was mixed with ammonium fluoride and heated to a temperature of 200°C. The interaction of phenakite with ammonium fluoride took place, with the formation of ammonium fluoroberrylate with ammonium hexafluorosilicate and the release of gaseous ammonia and water (24).

$$Be_2SiO_4 + 14NH_4F \rightarrow 2(NH_4)_2BeF_4 + (NH_4)_2SiF_6 + 8NH_3 + 4H_2O \tag{24}$$

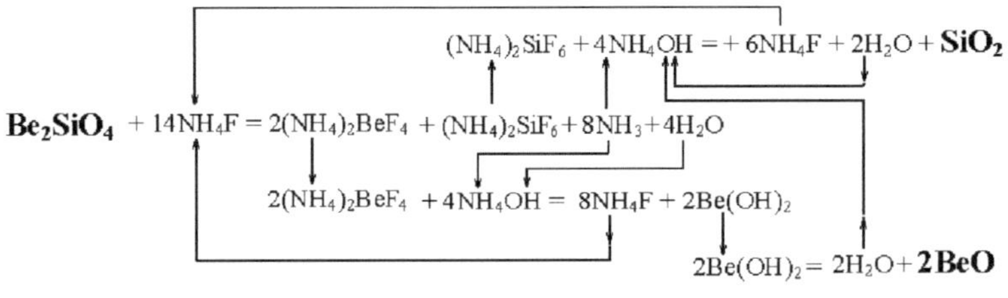

Figure 8.
Scheme of the decomposition of phenakite to beryllium oxide and silicon oxide.

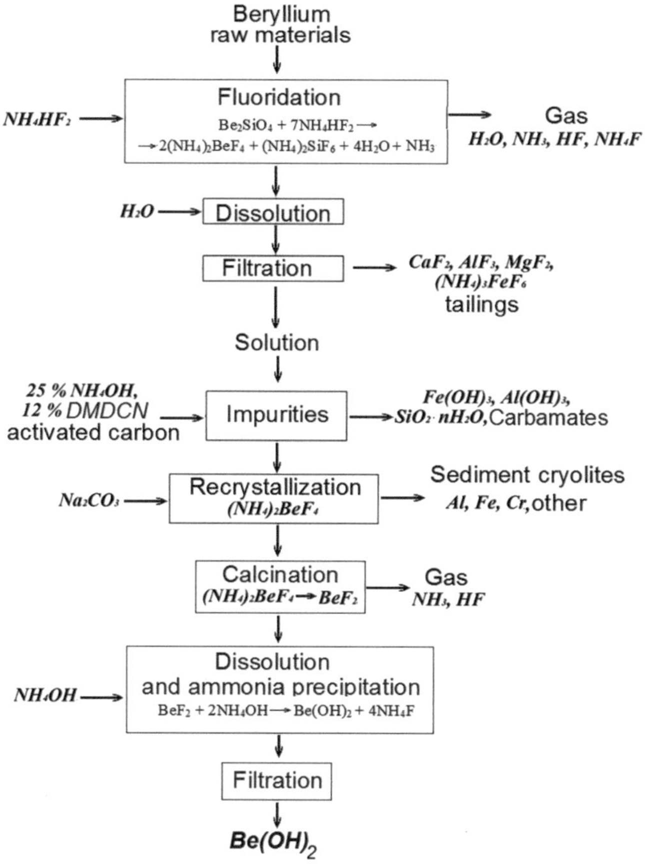

Figure 9.
Scheme of processing beryllium phenakite concentrate.

Upon heating, silicon was sublimated in the form of gaseous ammonium hexafluorosilicate $(NH_4)_2SiF_6$. The solid fraction contains ammonium tetrafluoro-berrylate and non-fluorinated fluoride. Leaching of ammonium tetrafluoroberrylate with water makes it possible to completely isolate it from the support rock, since the solubility of ammonium tetrafluoroberyllate reaches 32%, and calcium fluoride is practically insoluble. After separation of the ammonium tetrafluoroberyllate solution, it was purified from aluminum and iron impurities. Cleaning is carried out by the method of ammonia raising the pH of the solution to 8.5. At pH = 8.5, aluminum and iron hydroxides precipitate from the solution.

The separation of aluminum impurities by means of ammonia precipitation makes it possible to apply the fluoroammonium method also to the processing of beryl ($2BeO{\bullet}Al_2O_3{\bullet}6SiO_2$). With a further increase in pH to pH = 12, a precipitate of beryllium hydroxide precipitates. The beryllium hydroxide separated by filtration, after calcination, transforms into the oxide form.

A technological scheme of fluorine-ammonium processing of fluorite-phenakite concentrate with the return of ammonium fluoride to the process and the release of beryllium oxide, silicon oxide and calcium fluoride is proposed [15].

4. Conclusion

The optimum conditions of natural quartz processing were determined. Sublimation purification of $(NH_4)_2SiF_6$ is carried out at temperatures from 280 to 380°C and desublimation at 110–120°C. Synthetic silicon oxide with basic substance SiO_2 content of 99.999% was obtained.

Technological scheme for fluoroammonium processing of zirconium and obtaining zirconium dioxide is developed. Zircon is completely decomposed by ammonium hydrodifluoride at temperature 400°C and pressure 40 atm in an autoclave.

The technological scheme of ilmenite concentrate processing by ammonium fluoride to pigmented titanium dioxide and iron oxide (III) is developed. The optimal technological modes of ilmenite fluorination at 150–200°C for 2 hours are determined.

The methods for fluoroammonium processing of fluorite-phenakite concentrate and production of beryllium oxide, silicon oxide, and calcium fluoride were proposed. Fluorination occurs at 200°C, purification from silicon at 300–350°C.

We experimentally proved the feasibility of fluoroammonium processing of quartz, zircon, ilmenite, and phenakite minerals. The next step of our research we can propose the adaptation of fluoroammonium technologies for implementation in production.

Author details

Alexander Dyachenko
MIREA—Russian Technological University, Moscow, Russia

*Address all correspondence to: dyachenko@mirea.ru

References

[1] Nakajima T, Zemva B, Tressaud A. Advanced Inorganic Fluorides: Synthesis, Characterization and Applications. 1st ed 2000, eBook. Elsevier Science S.A.; 2000. ISBN: 9780080525488

[2] ASTM E 473-00. Standard definition of terms relating to thermal analysis. 2000

[3] Rakov EG, Mel'nichenko EI. The properties and reactions of ammonium fluorides. Russian Chemical Reviews. 1984;**53**(9):851-869. DOI: 10.1070/RC1984v053n09ABEH003126

[4] Tyagi AK. On the reaction of ammonium hydrogen fluoride with aluminum, nickel, chromium and zircaloy. Synthesis and Reactivity in Inorganic and Metal-Organic Chemistry. 1996;**26**(1):139-146. DOI: 10.1080/00945719608004252

[5] Rimkevich VS, Pushkin AA, Malovitskii YN, et al. Fluoride processing of non-bauxite ores. Russian Journal of Applied Chemistry. 2009;**82**: 6-11. DOI: 10.1134/S1070427209010029

[6] Habashi F. Kinetics of Metallurgical Processes. Sainte Foy, Québec City: Métallurgie Extractive Québec; 1999

[7] Gelmboldt VO, Kravtsov VC, Fonari MS. Ammonium hexafluoridsilicates: Synthesis, structures, properties, applications. Journal of Fluorine Chemistry. 2019;**221**:91-102. DOI: 10.1016/J.JFLUCHEM.2019.04.005

[8] Kurilenko LN, Laptash NM, Merkulov EB, Gluschenko VY. About fluorination of silicon containing minerals by ammonium hydrodifluoride. Researched in Russia. 2002;**130**(021011): 1465-1471

[9] Laptash N, Maslennikova I. Hydrofluoride decomposition of natural materials including zirconium-containing minerals. IOP Conference Series: Materials Science and Engineering. 2016;**112**(1):012024. DOI: 10.1088/1757-899X/112/1/012024

[10] Wang Y, Zhang Y, Zhao X, Liang G. Fabrication and properties of amorphous silica particles by fluorination of zircon using ammonium bifluoride. Journal of Fluorine Chemistry. 2020;**232**:109467. DOI: 10.1016/j.jfluchem.2020.109467

[11] Laptash N, Maslennikova I. Fluoride processing of titanium-containing minerals. Advances in Materials Physics and Chemistry. 2012;**2**(4B):21-24. DOI: 10.4236/ampc.2012.24B006

[12] Schmidbauer H. Recent contributions to the aqueous coordination chemistry of beryllium. Coordination Chemistry Reviews. 2001; **215**:223-242. DOI: 10.1016/S0010-8545(00)00406-9

[13] Thorat DD, Tripathi BM, Sathiyamoorthy D. Extraction of beryllium from Indian beryl by ammonium hydrofluoride. Hydrometallurgy. 2011;**109**:18-22. DOI: 10.4172/2168-9806.1000118

[14] Zaki EE, Ismail ZH, Daoud JA, Aly HF. Extraction equilibrium of beryllium and aluminium and recovery of beryllium from Egyptian beryl solution using CYANEX 921. Hydrometallurgy. 2005;**80**(4):221-231. DOI: 10.1016/j.hydromet.2005.07.009

[15] Andreev AA, Dyachenko AN, Kraidenko RI. Fluorination of beryllium concentrates with ammonium fluorides. Russian Journal of Applied Chemistry. 2008;**81**:178-182. DOI: 10.1134/S1070427208020043